Fundamentals of Propulsion

Fundamentals of **Propulsion**

Aryandra K. Jouhari
Former Professor of Mechanical Engineering, ASET and
Former Chief Scientist,
CSIR-Institute of Minerals and Materials Technology,
Bhubaneswar-751013, India
&
Sumit Krishnan
Assistant Professor of Aerospace Engineering, ASET
Amity University Uttar Pradesh, Lucknow Campus,
Lucknow-226028, India

ALLIED PUBLISHERS PVT. LTD.

New Delhi • Mumbai • Kolkata • Chennai • Bangalore • Hyderabad

ALLIED PUBLISHERS PRIVATE LIMITED

D-5, Sector-2, **Noida**–201 301
Ph. Nos.: 0120-4320295/2542557/4352866 • E-mail: delhi.books@alliedpublishers.com

17 Chittaranjan Avenue, **Kolkata**–700072
Ph.: 033-22129618 • E-mail: cal.books@alliedpublishers.com

15 J.N. Heredia Marg, Ballard Estate, **Mumbai**–400001
Ph.: 022-42126969 • E-mail: mumbai.books@alliedpublishers.com

No. 25/10, Commander-in-Chief Road, Ethiraj Lane (Next to Post Office)
Egmore, **Chennai**–600008
Ph.: 044-28223938 • E-mail: chennai.books@alliedpublishers.com

P.B. No. 9932, No. 15, 3rd Floor (Next to Vijaya Bank), 5th Cross, Gandhinagar, Karnataka, **Bangalore**–560009, Ph.: 080-41530285 / 22386239
• E-mail: bngl.journals@alliedpublishers.com / apsabng@airtelmail.in

Sri Jayalakshmi Nilayam, No. 3-4-510, 3rd Floor (Above More Super Market)
Barkatpura, **Hyderabad**–500027
Ph.: 040-27551811, 040-27551812 • E-mail: hyd.books@alliedpublishers.com

Website: www.alliedpublishers.com

ISBN: 978-93-90951-49-9

Published by Sunil Sachdev and printed by Ravi Sachdev at Allied Publishers Pvt. Ltd.,
D-5, Sector-2, Noida–201 301

The book is

Dedicated

to

the memory of

Chokhe Lal Satyapal

Vidyavati

Parents of Aryandra Kumar Jouhari

&

Ajay Kumar Srivastava

Father of Sumit Krishnan

"If he (the teacher) is wise
he does not bid you to enter the house of his wisdom,
but leads you to the threshold of your own mind."

—Gibran
The Prophet

Foreword

I am happy to write this Foreword for the textbook titled "Fundamentals of Propulsion" authored by Aryandra K. Jouhari and Sumit Krishnan. I have known the senior author (AKJ) for over thirty years now. For the first ten years, when I was in the Institute of Minerals and Materials Technology, Bhubaneswar, he was a valuable colleague and a friend. During my stay, he did his doctoral work under the guidance of Professor R.C. Gupta of Banaras Hindu University (BHU) and myself and submitted an excellent thesis in BHU to receive a well-deserved Ph.D. degree. He wrote several papers based on his thesis and they were published in reputed journals. Unfortunately, I did not have the opportunity of knowing Sumit Krishnan. After his retirement, AKJ joined the world of academics and became a Professor of Mechanical Engineering at Amity School of Engineering and Technology (ASET) in Lucknow. I am extremely happy that he has chosen to write this book with Sumit Krishnan as co-author, based on his lecture notes drawing material from many sources. The authors have used the COVID-19 lockdown period very well indeed. There is a very beautiful quotation from the writings of Kahlil Gibran in the beginning of the book which is very apt.

The book has several chapters that are all well written in simple language. These are: Introduction, Heat Transfer, Propeller Aerodynamics, Combustion, Internal Combustion Engines, Gas Turbines, Gas Dynamics, Intake and Propelling Nozzle, Ideal Jet Engine Cycle Analysis, Real Jet Engine Cycle Analysis, Axial Flow Compressor, Axial Flow Turbine, and an Appendix. Each chapter has appropriate sections and subsections, examples, a summary and a list of questions. The material presented in the book is meant to cover teaching for two semesters.

The text reads very well and has numerous figures and illustrations that enrich the book. There is a good mathematical treatment of topics and ideas presented in the book. I am sure that the students will find the book stimulating.

In the present age, the importance of Aerospace Engineering need not be over emphasized. The aerospace industry of aircraft, missiles, rockets, space vehicles, unmanned aerial vehicles etc. all exploit propulsion. The engineering principles of propulsion are used in other areas as well. This book is a very good addition to the literature on the subject.

I am sure that undergraduate students of this discipline will be immensely benefited by this book. The book will also be useful to graduate students and working professionals. It will be useful, as the authors say, for students preparing for the Graduate Aptitude Test in Engineering (GATE).

I congratulate the authors for writing this excellent textbook on an important subject.

Kolkata
4th January 2023

Professor Hem Shanker Ray
Distinguished Alumnus,
Indian Institute of Technology, Kharagpur,
Formerly, Director, Regional Research Laboratory,
(CSIR-Institute of Minerals and Materials Technology),
Bhubaneswar–751013, India.

Preface

B. Tech (Aerospace Engineering) course at Amity School of Engineering and Technology (ASET), Amity University Uttar Pradesh (AUUP), Lucknow Campus was introduced in the year 2011 with an intake of 19 students and the principal author has taught Propulsion I and II to semesters IV and V students respectively from the beginning. This course begins with heat transfer followed by propeller aerodynamics, combustion, internal combustion engines and gas turbines. Propulsion II in semester V begins with gas dynamics, followed by intake and propelling nozzle, thermodynamic analysis of ideal and real jet engines, axial flow compressor and axial flow turbine. The lecture notes prepared have been rewritten in the form of a book by the authors. Prior knowledge of Mathematics, Thermodynamics and Fluid Mechanics is expected from the reader.

We hope the book will be of great help to the students of Aerospace Engineering of not only Amity University but also to the students at other universities as well. Effort is made by the authors to provide the full course content of the university at one place. For further reading, advanced books may be referred, and the list is given at the end.

The authors would like to put on record the encouragement, support and discussions held with Professors H.S. Ray, former Director, Regional Research Laboratory (renamed as CSIR-Institute of Minerals and Materials Technology), Bhubaneswar, R.C. Gupta, former Professor and Head, Department of Metallurgical Engineering, IIT-Banaras Hindu University, Varanasi, Job Kurian, Former Professor and Head, Department of Aerospace Engineering and Dean of Industrial Consultancy & Sponsored Research, IIT Madras and Professor of Mechanical Engineering, IIT Palakkad, Late S.T.H. Abidi, M.A. Murtaza, Vinod Saxena and Professor (Dr.) O.P. Singh, Director, ASET, Lucknow Campus. Thanks are also due to Dr. Maneesh C. Srivastava, Head, Department of Mechanical and Aerospace Engineering, ASET, Col. Vivek Mathur, Head, Department of Civil Engineering, ASET and Assistant Professors Vivek Verma, Ashwani Sharma, Mragank Sharma and Dr. Kheelraj Pandey, ASET for going through the manuscript and giving useful suggestions.

The authors hope that the reader will enjoy the pages ahead and have some insight into the ocean of Propulsion.

Jai Hind...

Makar Sankranti

14th January 2023
Lucknow

Aryandra K. Jouhari
Sumit Krishnan

Acknowledgement

The authors initiated the writing of the book during lockdown due to Corona Virus (COVID-19) about two and a half years ago. The study material has been drawn from more than twenty books and the list is given in bibliography. The authors would like to express their gratitude to all the authors of the books and their publishers and thank them.

Most of the numerical examples and the questions given in each chapter are based on the end semester examination question papers of Amity University Uttar Pradesh. A few questions from GATE examination have also been incorporated.

The authors would like to express their gratitude to Founder President, Ritnand Balved Education Foundation, Dr. Ashok K. Chauhan and Chairman Lucknow Campus and Chancellor Amity University, C-VI Dr. Aseem K. Chauhan for giving them opportunity to work at Amity University Lucknow Campus. Thanks are also due to Wg. Cdr. (Dr.) Anil Kumar, Deputy Pro-Vice Chancellor, Amity University Uttar Pradesh, Lucknow Campus for his support and co-operation. The authors are grateful to Professor Hem Shanker Ray for going through the manuscript and writing the Foreword.

Last but not the least, we would like to express our gratitude to Late Ajay Kumar Srivastava and Shrimati Sunita, parents of Sumit Krishnan for encouraging him to undertake the work of writing the book. Thanks, are also due to Nidhi, wife of Sumit Krishnan for her support in preparing the manuscript. The contributions of Shrimati Sushma Jouhari and daughters Sarika, Meenakshi and Rashmi during preparation of the text are gratefully acknowledged.

Aryandra K. Jouhari
Sumit Krishnan

Contents

List of Symbols

A	area
A_j	area of nozzle
A_w	wetted area
A^*	area corresponding to sonic flow
a	velocity of sound, acceleration, constant, area
B	by-pass ratio
b	brake horsepower, constant
bp	brake power
C	specific heat, coefficient
C_p	specific heat at constant pressure
C_{pa}	specific heat at constant pressure for air
C_{pg}	specific heat at constant pressure for products of combustion
C_{pm}	specific heat at constant pressure for mixture of hot and fan stream
C_h	work output coefficient of hot/core stream
C_{prop}	work output coefficient of propeller
C_{total}	work output coefficient of turboprop
c	absolute velocity of fluid
$\bar{c}$	velocity vector
c_a	axial component of absolute velocity
c_r	radial component of absolute velocity
c_w	whirl/tangential component of absolute velocity
D	drag, diameter, hydraulic diameter
d	diameter
E	energy, modulus of elasticity, enrichment
e_c	polytropic efficiency of compressor
e_t	polytropic efficiency of turbine
F	force, drag, fan stream, thrust, impulse function
F_h	thrust produced in hot stream
F_{prop}	propeller thrust
F_s	thrust per unit mass flow rate of air
f	fuel/air ratio

fp	friction power
g	gravitational constant
g_c	Newton's constant, proportionality constant
h	enthalpy per unit mass, convection heat transfer coefficient
h_a	enthalpy of ambient air
h_0	total enthalpy
h_{PR}	lower heating value of fuel
ip	indicated power
J	advance ratio
K	bulk modulus, constant, number of cylinders
k	thermal conductivity
k_o	base thermal conductivity
L	length, lift
M	Mach number, mass
m	mass
m_a	mass of air
m_f	mass of fuel
m_j	mass of fluid in jet
$\dot{m}$	mass flow rate
$\dot{m}_a$	mass flow rate of air
$\dot{m}_F$	mass flow rate of air through fan stream
$\dot{m}_f$	mass flow rate of fuel
$\dot{m}_{f\,AB}$	mass flow rate of fuel in afterburner
$\dot{m}_h$	mass flow rate of gases in hot/core stream
$\dot{m}_j$	mass flow rate of gases in jet
$\dot{m}_{jh}$	mass flow rate of gases in jet of hot stream
$\dot{m}_{prop}$	mass flow rate of air in propeller
N	revolutions per minute/second
n	number of power stroke per minute
P	shaft brake power, power input
P_A	power available to aircraft
P_{if}	power input factor
p	absolute pressure
p_a	ambient pressure
p_c	pressure of choked flow

p_e	exit pressure
p_j	pressure of jet
p_o	stagnation / total pressure
p_{im}	indicated mean effective pressure
p_m	mean effective pressure
Q	heat
Q_R	heat rejected
Q_S	heat supplied
q	heat transfer
$\dot{q}$	heat transfer rate
q_{gen}	energy generated per unit volume
q_x, q_y, q_z	heat transfer in x, y, and z direction respectively
q_r	heat transfer in radial direction
R	gas constant
R_a	gas constant for air
R_g	gas constant for products of combustion
R_{th}	thermal resistance
r	radius, pressure ratio, compression ratio
r_c	cut-off ratio
r_p	pressure ratio
S	thrust specific fuel consumption
s	entropy per unit mass
T	absolute temperature, temperature, thrust, torque
T_a	ambient temperature
T_∞	free stream temperature
T_o	stagnation/total temperature
T_w	wall temperature
t	temperature ratio
U	total internal energy
u	internal energy per unit mass, peripheral velocity
V	velocity, relative velocity, volume
V_a	velocity of ambient air, flight speed
V_j	jet velocity
V_o	upstream air velocity
V_∞	freestream velocity

V_c	clearance volume
V_s	swept volume
V_{tip}	velocity at tip of propeller
$\forall$	volume
V_{jF}	jet velocity of fan stream
V_{jh}	jet velocity of hot / core stream
V^*	velocity at sonic condition (M=1)
$\bar{V}$	velocity vector
v	specific volume
W	work
$\dot{W}$	power
W_c	compressor work
W_E	engine work
W_N	net work
W_T	turbine work
$\dot{W}_{prop}$	propeller power input
x, y, z	cartesian co-ordinates

Greek

α	angle of attack, fraction, thermal diffusivity, constant
β	shock angle, pitch angle, blade angle, constant
γ	ratio of specific heats
δ	boundary layer thickness, incremental change
Δ	incremental change
η	efficiency
η_b	burner efficiency
η_c	compressor efficiency
η_{diesel}	thermal efficiency of Diesel cycle
η_g	efficiency factor
η_i	isentropic efficiency
η_{ith}	indicated thermal efficiency
η_j	jet efficiency
η_m	mechanical efficiency
η_{mL}	mechanical efficiency of transmission for low pressure turbine

η_n	nozzle efficiency
η_o	overall efficiency
η_{otto}	thermal efficiency of Otto cycle
η_P	propulsive efficiency
η_r	ram efficiency
η_T	thermal efficiency
θ	angle
λ	inverse of equivalence ratio, work done factor
μ	Mach angle, dynamic viscosity
ν	kinematic viscosity
π	pressure ratio
ρ	density, fraction
ρ^*	density at M=1
σ	Stefan Boltzmann constant, stress, slip factor
τ	temperature ratio (exit to entry of the component), shear stress
τ_λ	enthalpy ratio
φ	angle, equivalence ratio, flow coefficient
ψ	blade loading coefficient/temperature coefficient
ω	angular speed

Subscript

a	air, atmosphere, ambient
avg	average
AB	afterburner
b	burner efficiency, blade material
c	compressor
ct	centrifugal tensile
d	diffuser, inlet
$diesel$	Diesel cycle
dry	afterburner not operating
i	stage
if	input factor
o	calibrated conditions
$otto$	Otto cycle

F	fan stream
f	fuel
f_{AB}	fuel in afterburner
g	efficiency factor, products of combustion
h	hot/core stream
i	isentropic, stage
j	jet
jF	fanstream jet
jh	hot stream jet
m	mechanical, mixture of hot and fan stream
mL	mechanical efficiency of low-pressure turbine
max	maximum
min	minimum
N	net (available)
n	nozzle, normal
n_1, n_2	components normal to the oblique shock
o	overall, stagnation/total
opt	optimum
$otto$	Otto cycle
p	propulsive
$prop$	propeller
R	heat rejected
r	ram, freestream, radius, pressure ratio, root
rev	reversible
S	heat supplied
s	specific, stage
T	thermal
t	turbine, tip
tot	total
w	wall, wetted, whirl
$a, 1, 2, \ldots\ldots.., 19$	different locations in space

Superscript

*	state corresponding to M=1

Abbreviation

BDC	bottom dead center
bhp	brake horsepower
CI	compression ignition engine
fp	friction power
HPT	high pressure turbine
hp	horsepower
ip	indicated power
$imep$	indicated mean effective pressure
LPT	low pressure turbine
SFC	specific fuel consumption
SI	spark ignition
TDC	top dead center
TP	thrust power

Dimensionless numbers

$M = \frac{V}{a}$	Mach number
$N_u = \frac{hx}{K}$	Nusselt number
$P_r = \frac{\mu C_p}{k}$	Prandtl number
$Re_x = \frac{V.d.\rho}{\mu}$	Reynolds number
$G_{rx} = \frac{g\beta(T_W - T_\infty)x^3}{\nu^2}$	Grashof number

CHAPTER 1

Introduction

1.1 DEFINITIONS

The word propel means "to drive, or cause to move forward or onward" and propulsion may be defined as "the act of propelling, the state of being propelled, or propelling force or an impulse". Turbo machines are the systems in which energy is transferred to or from flowing fluid to a set of moving blade rows. The word turbo is of Latin origin and means "spins or whirls around". In a moving blade both enthalpy and pressure of the fluid changes. When energy is added to the fluid, it is called a compressor and when energy is taken out of the fluid, it is a turbine.

1.2 HISTORICAL

In the history of aeronautics, the premier aeronautical engineers are the Wright brothers, Wilbur (1867–1912) and Orville (1871–1948), when they flew their first heavier than air machine, the Wright Flyer I, and took the first flight on December 17, 1903, Thursday, at Kill Devil Hills, Kitty Hawk, North Carolina, US. The machine had two wings, one placed above the other. A 12 hp engine was mounted on the top surface of the bottom wing. The machine rose to a height of 3 m. The flight lasted for 12 s and the machine moved to a distance of 36 m. The machine had two propellers driven by chain and pulley arrangements connected to the same engine. The younger brother Orville was in command and Wilbur ran on the right side of the machine. Figure 1.1 shows a view of the Wright Flyer I. The first flight photographs were taken by John Daniels. The model is preserved in the Air and Space Museum of the Smithsonian Institution in Washington.

Wright brothers have acknowledged the contributions of their predecessors from whom they have learnt the various aspects of flying. It is interesting to note that neither of the two brothers received a high school diploma. Some other notable contributors in the aeronautical area are Leonardo da Vinci for his contributions for designing ornithopters in the end of 15th century. In 1783, a balloon went in air in Paris. George Cayley has worked on fixed wing concept for generating lift. He has stated that the lift is generated by a region of low pressure on the surface of the wing. Only George Cayley contributions may be considered comparable to Wright brothers.

Figure 1.1: A view of the Wright Flyer I, 1903
(*Source:* Anderson, Jr. John D., Introduction to Flight, Seventh Edition, McGraw-Hill 2012)

In 1809, gliders were built and flown. Otto Lilienthal (1848–1896) is regarded as glider man, and he flew the first successful controlled gliders. In the US, pioneering efforts were made by Samuel Pierpont Langley (1834–1906). It is he who considered that gasoline fueled engine is the solution to propel the aircraft. Another aviation pioneer is Glenn H. Curtiss, who became known throughout the world by 1910. He has built airplanes which became more popular and easier to fly. Figure 1.2 gives photographs of the four giant pioneers in the field of aeronautics.

Figure 1.2: The Four Giants Forming an Aeronautical Triangle

1.3 DIMENSIONS AND UNITS

A dimension is a physical quantity which specifies the nature of a particular system. The fundamental dimensions are length, mass, and time. For example, the length of a rod is the dimension of the rod. Temperature is a thermodynamic dimension. Though force is a derived dimension, for all practical purposes it is treated as a fundamental dimension. The fundamental dimensions are:

Length = L, Mass = M, Time = t, Temperature = T and Force = F

Force is defined by Newton's second law of motion, which states that force is proportional to the rate of change of momentum. The force F,

$$F \propto m.a$$

where, m = mass, a = acceleration

Therefore,

$$F = \frac{m.a}{g_c}$$

where, g_c = constant of proportionality

A unit is a system to define the dimensions. In the present era, SI (Système International d'unités) system is in use. In the earlier days, FPS (Foot, Pound, Second), CGS (Centimeter, Gram, Second) and MKS (Meter, Kilogram, Second) systems were in use.

In SI system the units are,

Mass = kg (kilogram), Length = m (meter), Time = s (second),
Temperature = °C, K (Kelvin) and Force = N (Newton).

In SI system, the force F,

$$1\,N = \frac{1}{g_c}.1\,kg.1\,m/s^2$$

$$g_c = 1.\frac{kg.m}{N.s^2}$$

Therefore, in SI system: $F = m.a$, and the unit of force, $1N = 1\text{ kg. m/s}^2$

Temperature in Kelvin, $K = °\text{C} + 273.16$

The weight of a body is equal to a product of mass and acceleration due to gravity.

$$\text{Work} = \text{Force} \times \text{Distance} = 1\,N.m$$

Work and energy (heat) units Joule (J), are same

$$1\,J = 1\,N.m$$

Rate of doing work is designated as power.

$$\text{Power} = \text{Work/Time}$$

$$1\,J/s = 1\,Nm/s = 1\,W\ (\text{Watt})$$

The unit of thermal conductivity, $k = W/m.°C$,

and for convection heat-transfer coefficient, $h = W/m^2.°C$

The units of specific heat at constant pressure and at constant volume, C_p and $C_v = J/kg.°C$

$$\text{heat flux, } q/A = W/m^2$$

where, A is area.

1.4 THRUST EQUATION

Consider the propulsion system given in Figure 1.3. The air enters the system at a velocity V_a, pressure p_a. The intake air velocity is equal and opposite to the forward speed of the aircraft.

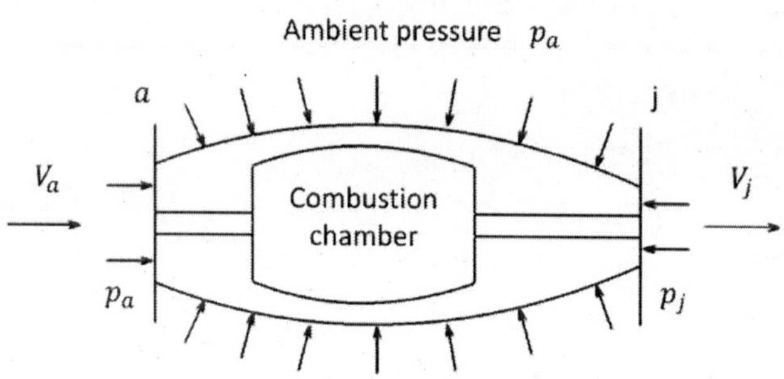

Figure 1.3: Propulsion System

Fuel is injected in the combustion chamber of the engine and the hot products of combustion at a velocity V_j, pressure p_j leave the propelling duct of the system. Let the mass flow of air intake be m_a and mass flow of fuel injected into the engine, m_f. The total mass of exit gases from the system, m_j will be,

$$m_j = m_a + m_f$$

The net thrust generated in the system is the sum of momentum thrust and pressure thrust. The momentum thrust is equal to the change in momentum between outlet gases and inlet air. The momentum is equal to the product of mass and velocity. Let A_j, be the area of the propelling nozzle. Then, the total thrust generated by the system is given as,

$$F = (m_j.V_j - m_a.V_a) + A_j(p_j - p_a)$$
$$= [(m_a + m_f).V_j - m_a.V_a] + A_j(p_j - p_a)$$

In the above equation, the first term is momentum thrust and the second term is pressure thrust.

For all practical purposes, $m_f \ll m_a$ and $p_j \approx p_a$ and hence the thrust generated will be:

$$F = m_a(V_j - V_a)$$

and the thrust power,

$$TP = F.V_a = m_a(V_j - V_a).V_a$$

When the aircraft is flying at a uniform speed, the thrust will be equal to drag.

The heat released by the fuel in combustion chamber will be,

$$= m_f.(h_{PR})$$

where, $h_{PR} =$ lower heating value of the fuel.

The chemical energy released by the fuel results in thrust. The change in kinetic energy of the system will be,

$$= \frac{1}{2}.(m_a + m_f)V_j^2 - \frac{1}{2}.m_a.V_a^2$$

As $m_f \ll m_a$, the kinetic energy change will be,

$$= \frac{1}{2}m_a(V_j^2 - V_a^2)$$

This gives useful thrust, and a part of energy is lost in the exit gases. The lost energy is equal to the kinetic energy of the jet relative to the earth and is equal to,

$$= \frac{1}{2}m_a(V_j - V_a)^2, \text{(as } m_f \ll m_a)$$

The propulsive efficiency,

$$\eta_p = \frac{\text{useful thrust power}}{\text{change in kinetic energy}}$$

$$= \frac{\text{useful thrust power}}{\text{useful thrust power + energy lost in the jet}}$$

$$= \frac{m_a.V_a(V_j - V_a)}{\frac{1}{2}m_a(V_j^2 - V_a^2)} = \frac{2.V_a}{V_j + V_a} = \frac{2}{1 + \left(\frac{V_j}{V_a}\right)}$$

It may be observed that η_p is maximum when $V_j = V_a$, but then the thrust generated will be zero and F is maximum when $V_a = 0$ but then $\eta_p = 0$. Therefore, for higher thrust, $V_j \gg V_a$ but then η_p will be less. Therefore, $V_j > V_a$ but the difference should be small to have better propulsion efficiency.

The efficiency of conversion of chemical energy of fuel, η_T is defined as,

$$\eta_T = \frac{\text{change in kinetic energy}}{\text{heat released from fuel}}$$

$$= \frac{\frac{1}{2}m_a(V_j{}^2 - V_a{}^2)}{m_f.(h_{PR})}$$

and the overall efficiency of the system,

$$\eta_o = \frac{\text{useful thrust power}}{\text{heat released from fuel}} = \eta_p.\eta_T$$

While calculating heat released from fuel, the burner efficiency may also be taken into consideration. In the case of turbojet engines, the burner efficiency is > 99.5%.

For evaluating the performance of a propulsion system, the term thrust specific fuel consumption, S is defined, which is equal to fuel consumption per unit thrust, and its unit is $kg/(N.s)$ in SI system.

Another important performance parameter is the specific thrust, F_S defined as thrust per unit mass flow rate of air, $N.s/kg$.

The fuel air ratio, $f = m_f/m_a$, the thrust specific fuel consumption, S and F_S are related as,

$$S = \frac{f}{F_S}$$

1.5 JET ENGINES

The schematic diagram of a simple gas turbine system is given in Figure 1.4. It consists of a compressor, a combustion chamber, and a turbine. The fuel is oxidized in combustion chamber and products of combustion at high temperature and high pressure are expanded in turbine to produce work. A part of the energy produced is used to run the compressor and the rest is expanded to produce shaft power.

In the case of a jet engine, the turbine produces shaft power just sufficient to operate the compressor and the rest of the hot gases are expanded in a propelling nozzle to produce thrust. Early gas turbines had low overall efficiency of about 40%. Over the years, with higher air pressure ratio in the compressor, 35:1; with 85–90% compressor and turbine efficiencies and turbine inlet gas temperature of 1650 K, very efficient gas turbines have been developed.

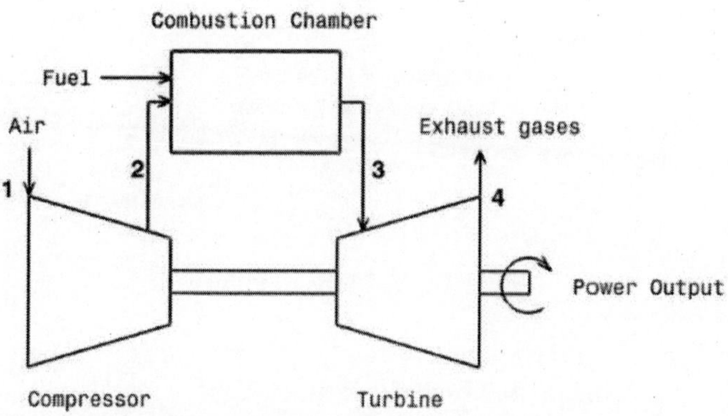

Figure 1.4: A Simple Gas Turbine System

In the previous section, it has been discussed that for higher propulsion efficiency, the difference between exit jet velocity and inlet air velocity should be small and, therefore, engines with large air mass flow rate and less difference in exit and inlet velocities were preferred. Accordingly, a family of jet engines has been developed. Initially, propellers were driven by piston engine and air was the medium of thrust. With the advent of gas turbine, in late 1930's, the propellers were driven by gas turbine and turbo-prop engines were developed. The turbojet engines are the engines with a compressor, combustion chamber, turbine, and a propelling nozzle. Then, turbofan engines came, where a part of the inlet air was by passed to produce thrust and rest of the required thrust was generated by the engine from the hot gases. These engines have the advantage of higher propulsion efficiency and lower noise as the exit gas velocity temperature was reduced. A ramjet engine has no compressor and therefore, no turbine. It has only a combustion chamber. These engines are most suited for flight Mach number in the range, 2.5–3.0, as in this range the specific fuel consumption is lowest. The air is compressed in an intake nozzle where the kinetic energy of air is converted to pressure energy. This phenomenon is designated as ram compression. The ramjet engines have the drawback that it can neither takeoff nor land without getting external support for both. Sometimes, an after burner is added in the engine, when additional thrust is required for a short duration, for example, in combat military operations. Figures 1.5–1.7 give diagrams of turbojet, turbojet with afterburner and turbofan engines. Specific thrust and thrust specific fuel consumption for various Mach numbers are given in Figures 1.8–1.9. Accordingly, proper selection of engine for a particular application is made.

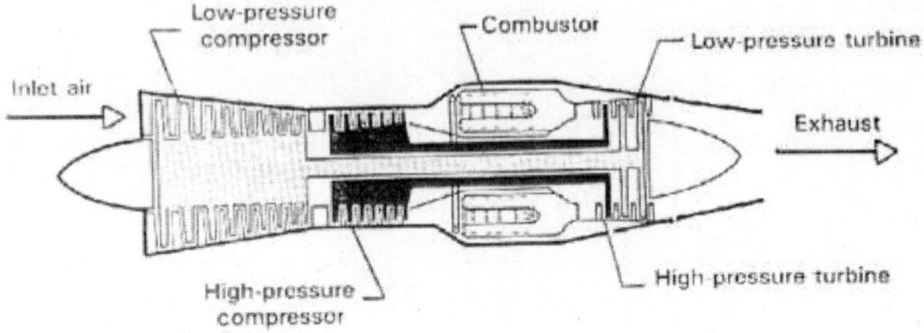

Figure 1.5: A Turbojet Engine

In the development of turbojet engines, the pioneers are: (i) Hans von Ohain, of Germany, who made the first successful flight with Heinkel engine on 27th August, 1939 and the other, (ii) Frank Whittle, of England, who made his first flight with a Gloster Whittle engine on 15th May, 1941. It may be noted that both have worked independently, without being aware of the work of each other.

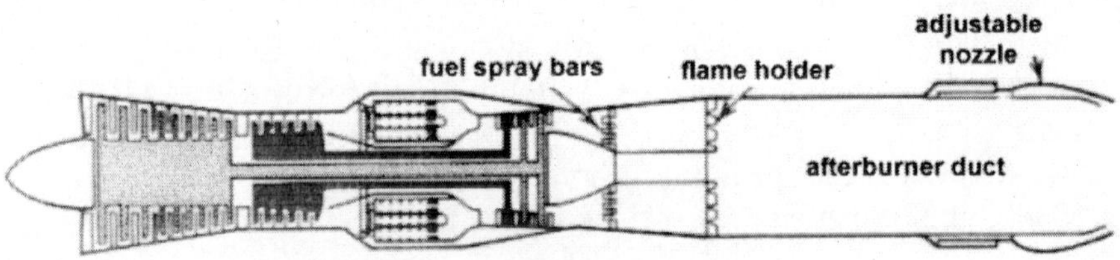

Figure 1.6: A Turbojet with Afterburner

Figure 1.7: A Turbofan Engine
(*Source:* Pratt & Whitney)

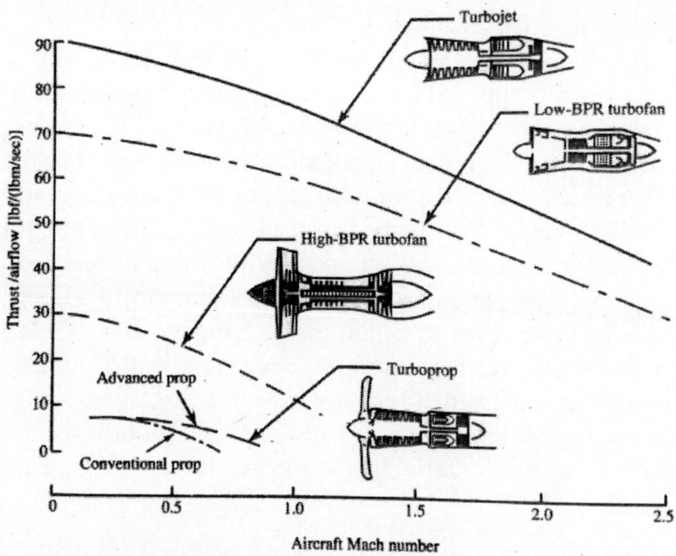

Figure 1.8: Specific Thrust Characteristics
(*Source:* Mattingly, Jack D., Elements of Gas Turbine Propulsion, McGraw Hill Education, 2017)

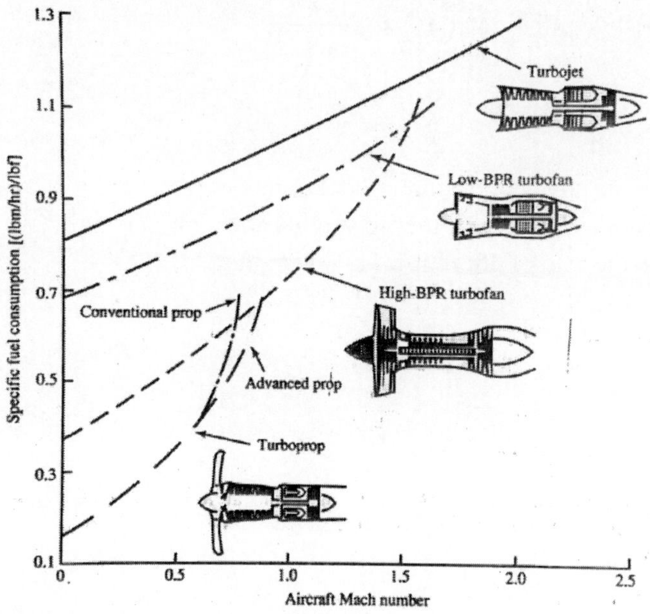

Figure 1.9: Thrust Specific Fuel Consumption Characteristics
(*Source:* Mattingly, Jack D., Elements of Gas Turbine Propulsion, McGraw Hill Education, 2017)

SUMMARY

In this chapter, we have defined the terms propel, propulsion and turbo machines. The word propel means to drive or move forward and propulsion may be defined as the act of propelling, the state of being propelled, or propelling force or an impulse. Turbomachines are rotary machines where energy is either added to the fluid or taken out of the fluid. When the energy is added, it is called a compressor and when the energy is taken out, it is called a turbine. The Wright brothers, Wilbur and Orville, are pioneers in the field of aeronautics. Some other names which need mention are Leonardo da Vinci, George Cayley, Otto Lilienthal, Samuel Langley and Glenn Curtiss. Dimensions and units are discussed next and units in S.I system for some important properties are given. The thrust equation, propulsion efficiency and overall efficiency of a jet engine are defined. In the last section, turbojet engines with and without afterburner, turbofan, turboprop, and ramjet have been introduced.

For jet engines,

$$\text{Thrust } (F) = m_a(V_j - V_a)$$

$$\text{Thrust Power } (TP) = m_a(V_j - V_a).V_a$$

$$\text{Propulsive efficiency } (\eta_p) = \frac{2}{1 + \left(\frac{V_j}{V_a}\right)}$$

QUESTIONS

1. Explain the advantages of a turbofan engine over a turbojet engine.
2. Differentiate between a turbojet engine and a ramjet engine.
3. Discuss the role of afterburner in jet engines.

CHAPTER 2

Heat Transfer

2.1 INTRODUCTION

Heat is a form of energy. While thermodynamics deals with the systems in equilibrium, heat transfer is the science which predicts the energy transfer rate between material bodies because of temperature difference. Temperature is defined by Zeroth law of thermodynamics, which states that, if two systems *A* and *B* are separately in thermal equilibrium with third system *C*, then systems *A* and *B* will also be in thermal equilibrium. This law gives the concept of temperature. Heat transfer is more an experimental subject. Heat flows from a higher temperature to lower temperature. There are three modes of heat transfer viz., conduction, convection, and radiation.

2.2 CONDUCTION

Consider a body, and the two ends of a body is having a temperature difference, then the heat flow will take place in the direction such that it is from a higher temperature to lower temperature. From experience it is known that heat transfer rate per unit area is proportional to the temperature gradient,

$$\frac{q}{A} \propto \frac{\partial T}{\partial x}$$

where, $q =$ heat transfer,

$A =$ Area,

$\frac{\partial T}{\partial x} =$ temperature gradient.

then,

$$\frac{q}{A} = -k\frac{\partial T}{\partial x}$$

where, $k =$ proportionality constant, designated as thermal conductivity.

The negative sign is since heat flows from higher to lower temperature.

The above equation is called the Fourier's law of heat conduction, named after French Physicist Joseph Fourier. The unit of k, thermal conductivity in SI system is $W/m\,.\,°C$.

Now, we will study the heat transfer by conduction in a solid, in cartesian co-ordinate system. Consider a one-dimensional system shown in Figure 2.1. The system is in steady state i.e., the temperature does not change with time and a heat source is present in the body. An element of thickness dx is drawn for analysis purpose.

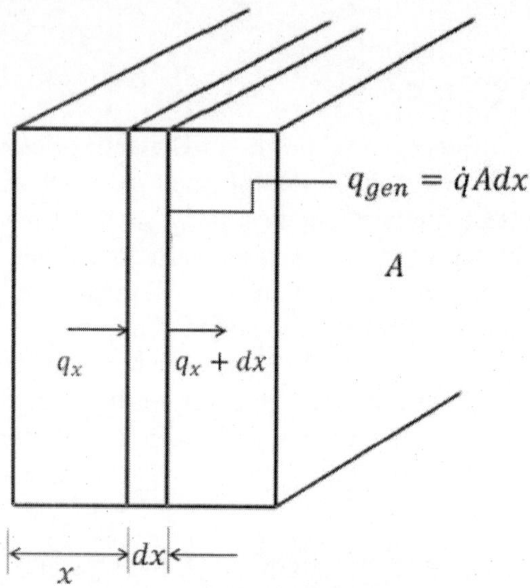

Figure 2.1: One Dimensional Heat Conduction in a Rectangular Block with Heat Source

(*Source:* Holman, J P and Bhattacharyya, S., Heat Transfer, Mc Graw Hill Education, Tenth Edition, 2016).

From law of conservation of energy,

$$\text{Energy on left face} + \text{energy generated in the element}$$
$$= \text{change in internal energy} + \text{energy conducted on right face}$$

Let, q_x be the heat flow in the x direction.

Heat flux on left face,

$$q_x = -kA\frac{\partial T}{\partial x}$$

Heat flux on right face,

$$q_{x+dx} = -kA\left(\frac{\partial T}{\partial x}\right)_{x+dx}$$

$$= -A\left[k\frac{\partial T}{\partial x} + \frac{\partial}{\partial x}\left(k\frac{\partial T}{\partial x}dx\right)\right]$$

Energy generated in the system due to heat source,

$$q_{gen} = \dot{q}Adx$$

where, $\dot{q}$ = energy generated/ volume, W/m^3

$$\text{Change in internal energy} = \rho.C.A\frac{\partial T}{\partial t}dx$$

where, C = specific heat of the material, J/kg °C

ρ = density, kg/m^3

t = time, s

According to energy equation,

$$-kA\frac{\partial T}{\partial x} + \dot{q}.A.dx = \rho CA\frac{\partial T}{\partial t}dx - A\left[k\frac{\partial T}{\partial x} + \frac{\partial}{\partial x}\left(k\frac{\partial T}{\partial x}\right)dx\right]$$

$$\frac{\partial}{\partial x}\left(k\frac{\partial T}{\partial x}\right) + \dot{q} = \rho C\frac{\partial T}{\partial t}$$

This is one dimensional heat conduction equation.

For three dimensional systems, we must consider the heat flux in y and z directions also. Then,

$$q_x + q_y + q_z + q_{gen} = q_{x+dx} + q_{y+dy} + q_{z+dz} + \frac{dE}{dt}$$

where, $\frac{dE}{dt}$ is change in internal energy with time.

Therefore, three-dimensional equation will be

$$\frac{\partial}{\partial x}\left(k\frac{\partial T}{\partial x}\right) + \frac{\partial}{\partial y}\left(k\frac{\partial T}{\partial y}\right) + \frac{\partial}{\partial z}\left(k\frac{\partial T}{\partial z}\right) + \dot{q} = \rho C\frac{\partial T}{\partial t}$$

If k is assumed constant, and divide by k,

$$\frac{\partial^2 T}{\partial x^2} + \frac{\partial^2 T}{\partial y^2} + \frac{\partial^2 T}{\partial z^2} + \frac{\dot{q}}{k} = \frac{\rho C}{k}\frac{\partial T}{\partial t}$$

In the above equation, the term $\frac{k}{\rho C}$ is designated as thermal diffusivity, α of the material.

Similarly, the conduction equations in cylindrical co-ordinates and spherical co-ordinates can be developed.

To solve practical problems, a few equations are as follows:

1. $\frac{\partial^2 T}{\partial x^2}$=0 (for steady state one-dimensional heat flow with no heat generation)

2. $\frac{d^2T}{dr^2} + \frac{1}{r}\frac{dT}{dr} = 0$ (for steady state one-dimensional heat flow with no heat generation in cylindrical co-ordinates)

3. $\frac{\partial^2 T}{\partial x^2} + \frac{\dot{q}}{k} = 0$ (for steady state one-dimensional heat flow with heat source)

4. $\frac{\partial^2 T}{\partial x^2} + \frac{\partial^2 T}{\partial y^2} = 0$ (for steady state two-dimensional heat flow without heat source)

2.2.1 Thermal Conductivity

Thermal conductivity is defined by the Fourier's law of heat conduction and is determined experimentally. In a few cases it is estimated theoretically. Thermal conductivity of a few materials is given in Table 2.1.

Table 2.1: Thermal Conductivity of Materials at 0°C

	Thermal Conductivity, $k, W/m.°C$
Metals	
Silver	410
Aluminium	202
Carbon Steel, 1% C	43
Non-metals	
Saw dust	0.059
Glass Wool	0.038
Liquids	
Mercury	8.21
Water	0.556
Gases	
Hydrogen	0.175
Air	0.024

(*Source:* Holman, J.P. and Bhattacharyya, S., Heat Transfer, Mc Graw Hill Education, Tenth Edition, 2016)

2.2.2 Plane Wall

Consider a plane wall shown in Figure 2.2(a).

The wall may be of one material or a composite with layers of many materials having different thermal conductivity.

For a wall made of one material,

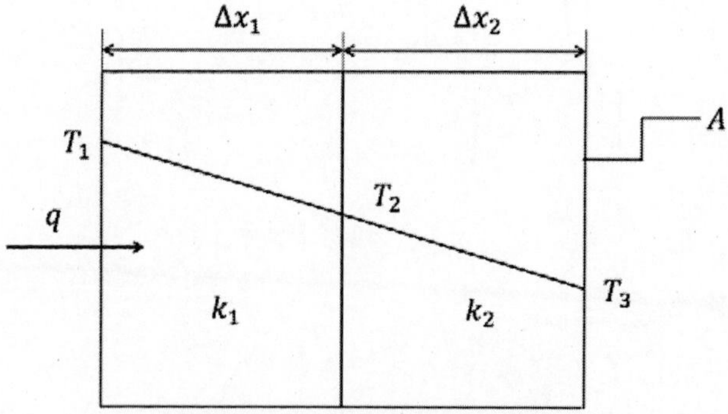

Figure 2.2(a): Composite Wall

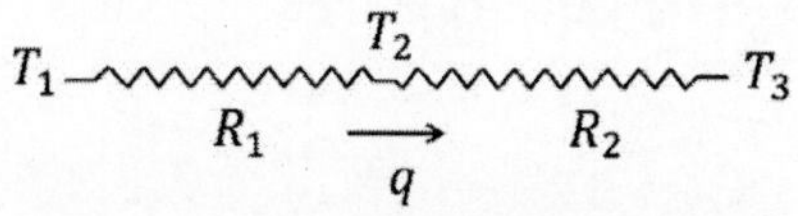

Figure 2.2(b): Electrical Analogy

According to Fourier's law,

$$q = -kA\frac{dT}{dx}$$

Integrate,

$$q\int dx = -kA\int_1^2 dT$$

$$q.\Delta x = -kA(T_2 - T_1)$$

$$q = -\frac{kA}{\Delta x}(T_2 - T_1)$$

Here, k is constant.

If k is a function of temperature, then

$$k = k_o(1 + \beta T)$$

where, k_o is base thermal conductivity and β is a constant. k_o is also constant.

Substitute in conduction equation and integrate,

$$q = -k_oA(1 + \beta T)\frac{dT}{dx}$$

$$qdx = -k_oA[dT + \beta TdT]$$

Integrate,

$$\int q\, dx = -k_o A \left[\int_{T1}^{T2} dT + \int_{T1}^{T2} \beta T\, dT \right]$$

$$q\Delta x = -k_o A \left[T + \beta \frac{T^2}{2} \right]_1^2$$

$$= -k_o A \left[(T_2 - T_1) + \frac{1}{2}\beta (T_2{}^2 - T_1{}^2) \right]$$

$$q = -\frac{k_o A}{\Delta x} \left[(T_2 - T_1) + \frac{1}{2}\beta (T_2{}^2 - T_1{}^2) \right]$$

For a composite plane wall, the temperature at interface is T_2 and at the ends are T_1 and T_3. $\Delta x_1$and Δx_2 are thicknesses of the two layers. Then, according to conduction equation (it may be noted that q remains constant),

$$q = -\frac{k_1 A}{\Delta x_1}(T_2 - T_1)$$

$$(T_2 - T_1) = -\frac{q\Delta x_1}{k_1 A}$$

$$(-T_2 + T_1) = \frac{q.\Delta x_1}{k_1 A}$$

Similarly,

$$q = -\frac{k_2 A}{\Delta x_2}.(T_3 - T_2)$$

$$(-T_3 + T_2) = \frac{q.\Delta x_2}{k_2 A}$$

Add the above two equations,

$$(T_1 - T_3) = \frac{q.\Delta x_1}{k_1 A} + \frac{q.\Delta x_2}{k_2 A}$$

$$= q\left[\frac{\Delta x_1}{k_1 A} + \frac{\Delta x_2}{k_2 A}\right]$$

$$q = \frac{(T_1 - T_3)}{\left[\frac{\Delta x_1}{k_1 A} + \frac{\Delta x_2}{k_2 A}\right]}$$

$$\text{Heat flow} = \frac{\text{thermal potential differences}}{\text{thermal resistance}}$$

$$R_1 = \frac{\Delta x_1}{k_1 A}, R_2 = \frac{\Delta x_2}{k_2 A}$$

$$q = \frac{(\Delta T)_{\text{overall}}}{\sum R_{th}}$$

$$\sum R_{th} = R_1 + R_2$$

where, R_1, R_2 are the thermal resistance.

This is called electrical analogy and is shown in Figure 2.2(b).

If the composite wall is a combination of materials as given in Figure 2.3(a), A and E in series and *B, C, D* and *F, G* in parallel. Then the electrical analogy will be as shown in Figure 2.3(b).

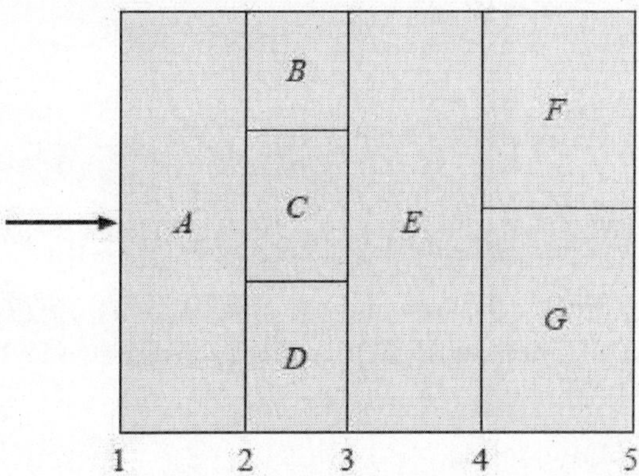

Figure 2.3(a): Series and Parallel 1D Heat Conduction Through a Composite Wall

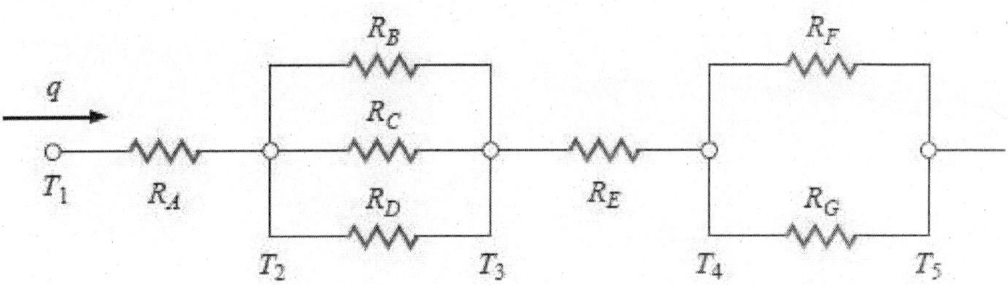

Figure 2.3(b): Electrical Analogy

(*Source:* Holman, J.P. and Bhattacharyya, S., Heat Transfer, Mc Graw Hill Education, Tenth Edition, 2016)

2.2.3 Cylinder

Consider a long hollow cylinder, inner radius r_i, outer radius r_o and length L (Figure 2.4).

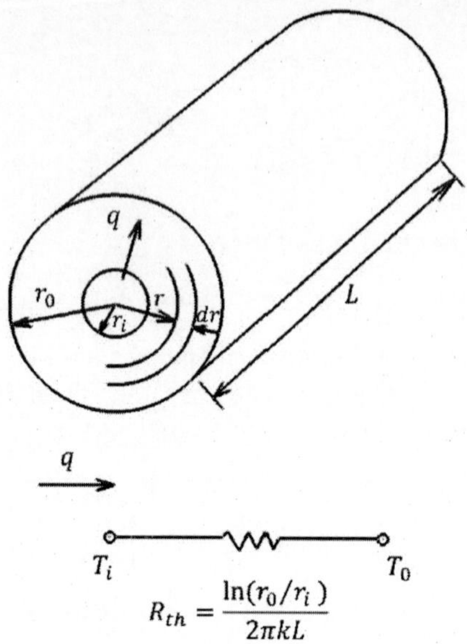

Figure 2.4: 1D Heat Flow Through a Hollow Cylinder and Electrical Analogy

The inner temperature is T_i and outer temperature is T_o ($T_i > T_o$). In one dimensional case, the heat will flow radially from inside to outside. For analysis, take a strip at radius r and length L.

Then, area for heat flow will be

$$A_r = 2\pi rL$$

Applying Fourier law, the heat flow at radius r

$$q_r = -kA_r \frac{dT}{dr}$$

$$q_r = -k2\pi rL \frac{dT}{dr}$$

The boundary conditions are,

$$\text{at}, r = r_i, T = T_i$$

$$\text{and}, r = r_0, T = T_0$$

$$q_r . \frac{dr}{r} = -2\pi kL\, dT$$

Integrate,

$$q \int_{r_i}^{r_o} \log_e r = -2\pi kL[T_o - T_i]$$

$$q.\log_e \frac{r_o}{r_i} = 2\pi kL[T_i - T_o]$$

$$q = \frac{2\pi kL[T_i - T_o]}{\log_e \frac{r_o}{r_i}}$$

By electrical analogy,

$$R_{th} = \frac{\log_e \frac{r_o}{r_i}}{2\pi kL}$$

And if the cylinder is having multiple layers, say of two materials having radius r_1, r_2, r_3. The thermal conductivity of material between radius r_1 and r_2 be k_1 and between r_2 and r_3 be k_2 then,

$$R_{th} = \frac{\log_e \frac{r_2}{r_1}}{2\pi k_1 L} + \frac{\log_e \frac{r_3}{r_2}}{2\pi k_2 L}$$

2.2.4 Sphere

In spherical system, heat flow is a function of radius only. Area of a section of sphere at radius r, is

$$A_r = 4\pi r^2$$

Applying Fourier law,

$$q_r = -k.4\pi r^2 \frac{dT}{dr}$$

$$q_r.\frac{dr}{r^2} = -4\pi kdT$$

Integrate,

$$q.\left[-\frac{1}{r}\right]_1^2 = -4\pi k(T_2 - T_1)$$

For a hollow sphere of inside radius r_1 and temperature T_1 and outside radius r_2, temperature T_2.

$$q\left[\frac{1}{r_1} - \frac{1}{r_2}\right] = 4\pi k(T_1 - T_2)$$

$$q = \frac{4\pi k(T_1 - T_2)}{\left(\frac{1}{r_1} - \frac{1}{r_2}\right)}$$

2.2.5 Plane Wall with Heat Source

Consider a plane wall with heat source as shown in Figure 2.5. Let the thickness of the plane wall be $2L$ and the distance x is measured from the center of the wall. Let both the wall temperatures be T_W and the heat generated per unit volume per unit time in the wall be $\dot{q}$. A nuclear reactor is an example of such a system. It is assumed that the thermal conductivity does not vary with temperature. Then, as discussed earlier, the differential equation for heat flow will be,

$$\frac{d^2T}{dx^2} + \frac{\dot{q}}{k} = 0$$

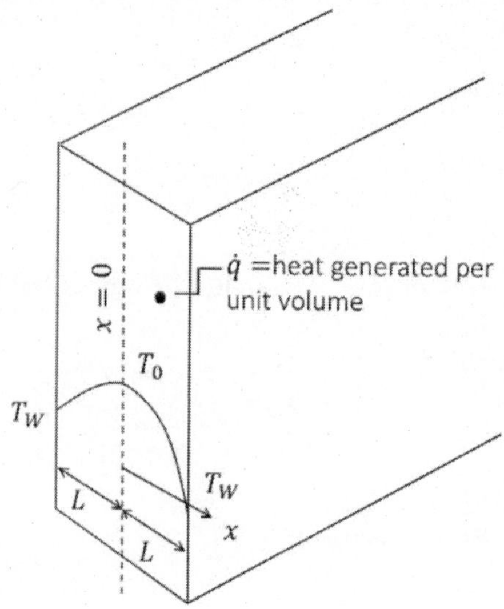

Figure 2.5: 1D Heat Conduction with Heat Generation

and the boundary conditions are,

$$T = T_W \text{ at } x = \pm L$$

The general solution for the equation will be of the form,

$$T = -\frac{\dot{q}}{2k}x^2 + c_1 x + c_2$$

$c_1 = 0$ because the temperature is same on both sides of the wall. Also, at $x = 0, T = T_0$.

therefore, $c_2 = T_0$

$$T - T_o = -\frac{\dot{q}}{2k}x^2$$

at $x = L, T = T_W$

$$T_w - T_o = -\frac{\dot{q}}{2k}L^2$$

$$\frac{T - T_o}{T_w - T_o} = \left(\frac{x}{L}\right)^2$$

This is a parabolic equation. Therefore, the temperature distribution of the wall will be parabolic as shown in Figure 2.5.

2.2.6 Two Dimensional—Steady State Heat Conduction

For steady state, two-dimensional heat flow with no heat generation, the equation is given as,

$$\frac{\partial^2 T}{\partial x^2} + \frac{\partial^2 T}{\partial y^2} = 0$$

for constant thermal conductivity. The solution of this equation will be of the form,

$$q_x = -kA_x \frac{\partial T}{\partial x}$$

$$q_y = -kA_y \frac{\partial T}{\partial y}$$

The heat flow quantities are either in the x-direction or in the y-direction. Figure 2.6 gives the heat flow in two dimensions and Figure 2.7 gives the isotherms and heat flow lines in a rectangular plate.

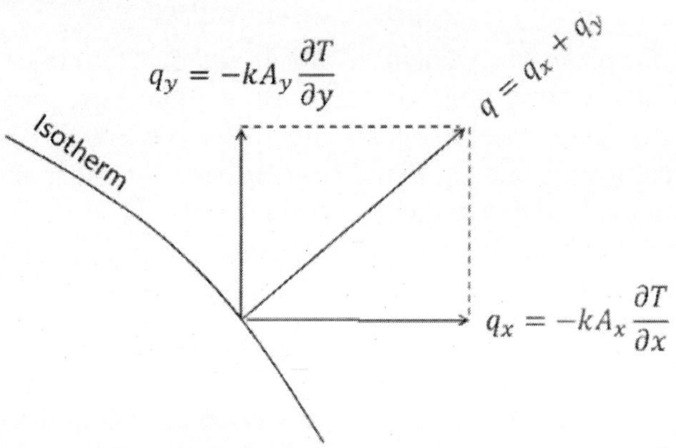

Figure 2.6: Heat Flow in Two Dimensions

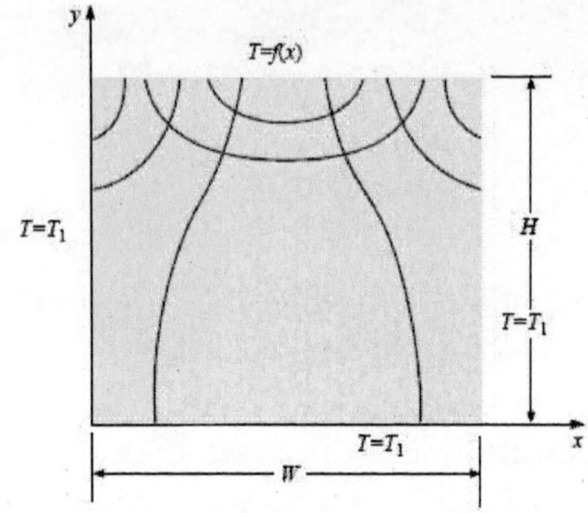

Figure 2.7: Isotherms and Heat Flow Lines in a Rectangular Plate

2.3 CONVECTION

Consider a plate at temperature, T_w as shown in Figure 2.8 and free air is flowing at a velocity,

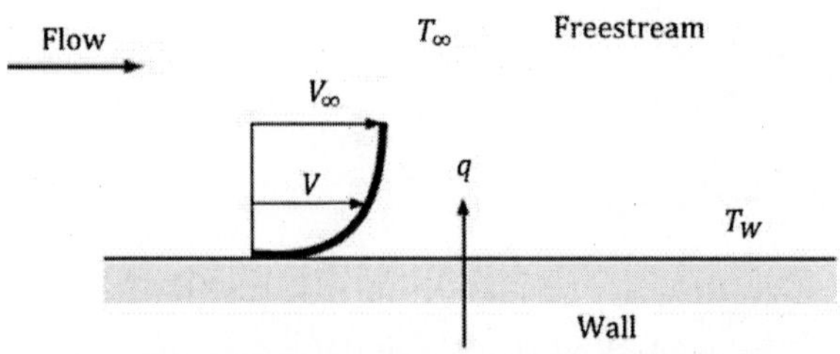

Figure 2.8: Convection Heat Transfer from a Plate
(*Source:* Holman, J P and Bhattacharyya, S., Heat Transfer, Mc Graw Hill Education, Tenth Edition, 2016)

V_∞ parallel to the plate. The velocity of the air at the plate surface will be zero and at some distance away from the plate, will be equal to free stream velocity. This thickness is called the boundary layer. The concept of boundary layer was first developed by Ludwig Prandtl, a German scientist in the year 1904. The exact solution to the boundary layer thickness was given by his student Blasius in the year 1908. This boundary layer thickness, δ is given as,

$$\delta = \frac{5.0}{\sqrt{\mathrm{Re_x}}}$$

where, $\mathrm{Re_x} = \frac{V_\infty . x . \rho}{\mu}$, Reynold's number at a distance x from the leading edge of the plate, ρ is density and μ is viscosity of the fluid. When the plate is heated at a temperature, T_w, the heat flux, q is given by,

$$q = hA(T_w - T_\infty)$$

where, T_∞ = free stream temperature, A = area of the plate and h = convection heat transfer coefficient, the unit of h in SI system is $\mathrm{W/m^2\,.\,°C}$

This phenomenon of heat flow is called convection.

Two more dimensionless numbers; Prandtl number (Pr) and Nusselt number (Nu) are defined as below:

$$Pr = \frac{\mu C_p}{k}$$

$$Nu = \frac{h.x}{k}$$

where, x is the characteristic dimension, k is thermal conductivity and C_p is specific heat at constant pressure of the fluid.

2.3.1 Thermal Boundary Layer

Consider a hot plate, Figure 2.9, where heating starts at $x = x_o$.

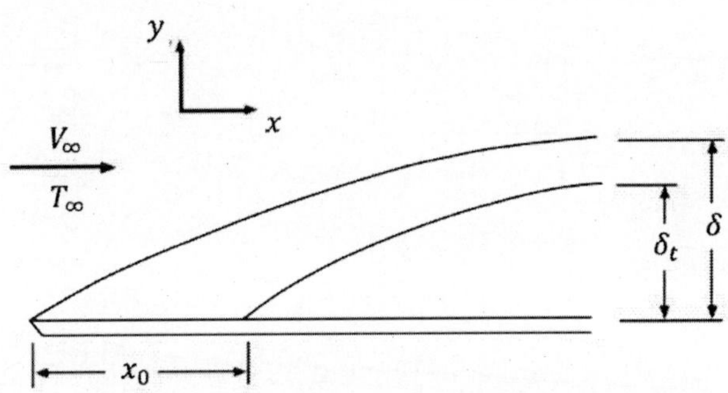

Figure 2.9: Flat Plate Heating Starts at $x = x_o$

In this case, two boundary layers are formed, one is called hydrodynamic boundary layer, thickness, δ and the other is called thermal boundary layer, thickness, δ_t.

When the plate is heated over its entire length, i.e. $x_o = 0$, then the relationship to calculate heat transfer coefficient, h_x at a distance x from the leading edge is given as:

$$Nu_x = 0.332\, Pr^{1/3}\, Re_x^{1/2}$$

where, Nu_x and Re_x are evaluated at a distance x from the leading edge.

The fluid properties are to be evaluated at film temperature, T_f, defined as:

$$T_f = \frac{T_w + T_\infty}{2}$$

The above analysis is valid for isothermal surface, T_w= constant. When the heat flux remains constant, h should be evaluated using the equation,

$$Nu_x = 0.453\, Pr^{1/3}\, Re_x^{1/2}$$

The above relations are valid under the following limitations:

$Re_x < 5 \times 10^5$ and

$0.6 < Pr < 50$

2.3.2 Pipe and Tube Flow

Consider a fluid flow in a tube or pipe, Figure 2.10.

Total energy, q added is given by

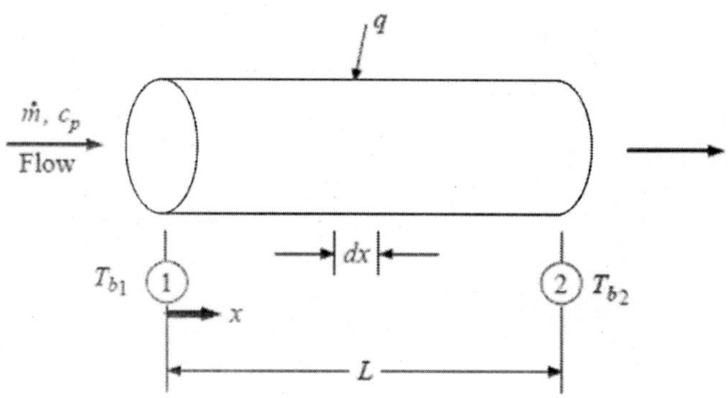

Figure 2.10: Heat Transfer in Terms of Bulk Temperature Difference

$$q = \dot{m} C_p (T_{b2} - T_{b1})$$

where, T_{b1} and T_{b2} = bulk temperature at exit and entry of the tube

The heat flux is radial, then heat flux dq at radius r for an element of length dx will be,

$$dq = \dot{m} C_p (dT_b)$$
$$= h(2\pi r)(T_w - T_b)$$

$\dot{m}$ = mass flow rate of fluid and the total heat transfer, q

$$q = hA(T_w - T_b)_{av}$$

where, $(T_w - T_b)_{av}$ is the average temperature difference.

To evaluate h, the relation for smooth tube is,

$$Nu_d = 0.023\, Re_d^{0.8} Pr^n$$

where, suffix d is for diameter, and

$n = 0.4$ for heating of fluid

$n = 0.3$ for cooling of fluid

and the properties should be evaluated at average fluid bulk temperature.

2.3.3 Natural Convection

An example is an electric radiator for room heating. The volume coefficient of expansion, β is $1/T$, where T is absolute temperature of the gas. A new dimensionless number Grashof number, G_{rx} is introduced as below.

$$G_{rx} = \frac{g\beta(T_w - T_\infty)x^3}{\nu^2}$$

where, ν is kinematic viscosity, $\frac{\mu}{\rho}$ and g is gravitational constant.

The heat transfer coefficient, h can be estimated using a relation,

$$\overline{N}_{uf} = c(G_{rf}.Pr_f)^m$$

where, $\overline{N}_{uf}$is average Nusselt number, c = constant, suffix f is far film temperature and m is a constant.

For laminar and turbulent flow,

$m = \frac{1}{4}$ for laminar

$m = \frac{1}{3}$ for turbulent

2.4 RADIATION

Thermal Radiation is defined as the electromagnetic radiation emitted by a body because of its temperature. An ideal thermal radiator is black body and the energy emitted is proportional to the fourth power of the absolute temperature and to surface area of the body.

$$q_{\text{emitted}} = \sigma A T^4$$

where, σ = Stefan – Boltzmann constant

$$= 5.669 \times 10^{-8} \frac{W}{m^2} \cdot \frac{1}{K^4}$$

The radiant heat exchange between two surfaces will be proportional to the difference in absolute temperature to the power four.

$$\frac{q}{A} \propto \sigma(T_1^4 - T_2^4)$$

and,

$$q = F_\epsilon F_G \sigma A(T_1^4 - T_2^4)$$

where, F_ϵ = emissivity function

F_G = geometric view factor

Consider a material surface, shown in Figure 2.11, and a radiant energy strike, then a part will be reflected, a part will be absorbed, and a part will be transmitted. ρ may be defined as the fraction reflected, α as the fraction absorbed and τ as the fraction transmitted. Then

$$\rho + \alpha + \tau = 1$$

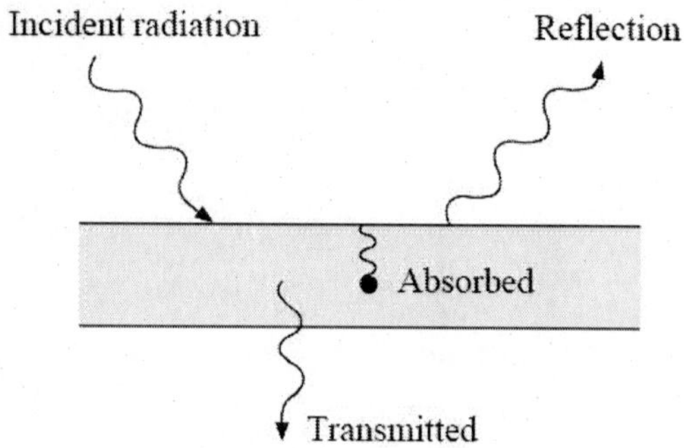

Figure 2.11: Effect of Incident Radiation on a Material Surface

From experience, it is known that most of the solid bodies do not transmit thermal radiation and therefore, $\tau = 0$. Then,

$$\rho + \alpha = 1$$

When the angle of incidence and the angle of reflection are same, the reflection is called specular and when the reflected beam gets distributed in all directions uniformly, the reflection is called diffuse.

A gray body is defined as a body which has monochromatic emissivity independent of wavelength. The monochromatic emissivity is the ratio of monochromatic emissive power of the body to the monochromatic emissivity of a black body at the same wavelength and temperature.

2.5 Combined Heat Transfer

Consider a plane wall shown in Figure 2.12, where hot fluid, A is on one side and the cold fluid, B is on the other side. Then the heat transfer, q.

$$q = h_1 A(T_A - T_1)$$

$$q = \frac{kA}{\Delta x}(T_1 - T_2)$$

$$q = h_2 A(T_2 - T_B)$$

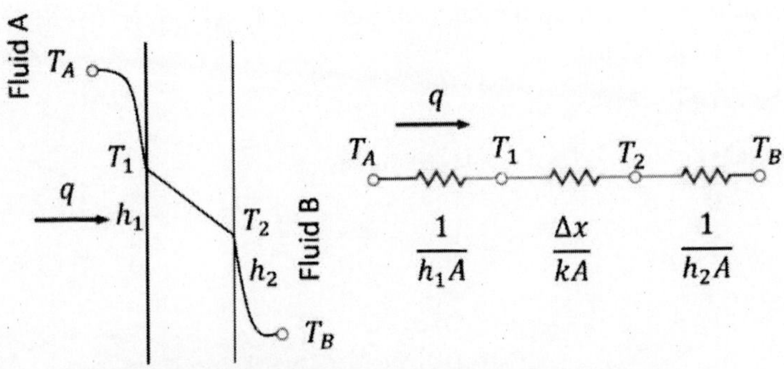

Figure 2.12: Plane Wall and Electrical Analogy

where, Δx is the wall thickness.

The overall heat transfer coefficient,

$$h_{\text{overall}} = \frac{1}{\left(\frac{1}{h_1} + \frac{\Delta x}{k} + \frac{1}{h_2}\right)}$$

Similarly, for a hollow cylinder with fluid flowing in both inside and outside cylinder, the overall heat transfer coefficient, h_{overall}

$$h_{\text{overall}} = \frac{1}{\frac{1}{h_1 A_1} + \frac{ln(r_2/r_1)}{2\pi kL} + \frac{1}{h_2 A_2}}$$

where A_1 and A_2 are the surface area of inside and outside of the tube where the fluid is flowing and r_1 and r_2 are inner and outer radius.

SUMMARY

In this chapter, the fundamentals of heat transfer have been discussed. Heat is a form of energy. Heat transfer deals with prediction of energy transfer rate between material bodies due to temperature difference. Temperature is defined by Zeroth law of thermodynamics. Heat flows from a higher temperature to lower temperature. Thermal conductivity, k is defined by the Fourier's law of heat conduction, viz.

$$\frac{q}{A} = -k\frac{\partial T}{\partial x}$$

The unit of k in S.I. system is $W/m.\,°C$.

Heat transfer by conduction is evaluated for a plane wall, cylinder, and sphere.

Convection heat transfer coefficient, h, unit in S.I. system $W/m^2.\,°C$, is given by heat flow equation as,

$$q = h.A.(T_w - T_\infty)$$

The boundary layer thickness, δ is given as,

$$\delta = \frac{5.0}{\sqrt{Re_x}}$$

and Reynolds number, Re_x

$$Re_x = \frac{V_\infty.x.\rho}{\mu}$$

For estimation of h, two dimensionless numbers, as below are defined,

Prandtl number, Pr

$$Pr = \frac{\mu C_p}{k}$$

and, Nusselt number, Nu

$$Nu = \frac{h.x}{k}$$

For an isothermal surface, the heat flux remains constant and is evaluated using the equation,

$$Nu_x = 0.332\, Pr^{1/3}\, Re_x^{1/2}$$

Thermal radiation is defined as the electromagnetic radiation emitted by a body as a result of its temperature. An ideal thermal radiator is black body and the energy emitted is proportional to the fourth power of the absolute temperature and to the surface area of the body.

$$q_{\text{emitted}} = \sigma A T^4$$

where, $\sigma = $ Stefan – Boltzmann constant

$$= 5.669 \times 10^{-8}\frac{W}{m^2}.\frac{1}{K^4}$$

EXAMPLES

Example 2.1

Evaluate the heat flux across a composite wall of width 0.30 m and 0.20 m of conductivity 385 $W/m.K$ and 202 $W/m.K$ respectively, when its one surface is at 200°C and the other surface is at 50°C. Also, calculate the temperature at the interface of the two materials of the wall.

Solution: Refer Figure 2.13

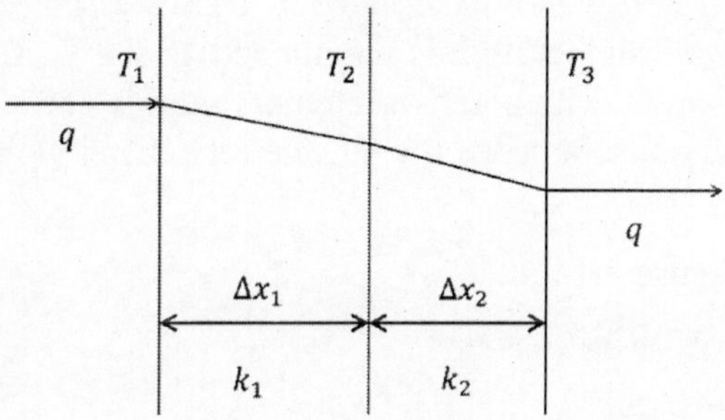

$T_1 = 200°C$

$T_3 = 50°C$

Figure 2.13 Heat flow through a plane composite wall

$\Delta x_1 = 0.30\ m$

$\Delta x_2 = 0.20\ m$

$k_1 = 385\ W/m.K$

$k_2 = 202\ W/m.K$

$$q = \frac{T_1 - T_3}{\left(\frac{\Delta x_1}{k_1 A} + \frac{\Delta x_2}{k_2 A}\right)}$$

$$\frac{q}{A} = \frac{T_1 - T_3}{\left(\frac{\Delta x_1}{k_1} + \frac{\Delta x_2}{k_2}\right)}$$

$$= \frac{(200 - 50)}{\left({}^{0.30}/_{385} + {}^{0.20}/_{202}\right)}$$

$$= \frac{150}{0.00078 + 0.00099}$$

$$= \mathbf{84778.34\ W}$$

$$\frac{q}{A} = \frac{T_1 - T_2}{(\Delta x_1/K_1)}$$

$$84778.34 = \frac{200 - T_2}{0.00078}$$

$$T_2 = \mathbf{133.87}\ ^\circ\mathrm{C}$$

Example 2.2

Consider a steel tube ($k = 43\ W/m.{}^\circ\mathrm{C}$) of inner radius 3 cm and thickness 2 cm. The tube is covered with a material ($k = 4.15\ W/m.{}^\circ\mathrm{C}$) having thickness 2 cm. the inside wall temperature is maintained at 800°C. Calculate the heat loss per meter of length. The outer wall temperature is 150°C. Also, calculate the interface temperature between the outside tube material and the inner tube material.

Solution: Refer Figure 2.14

$$r_1 = 3\ \mathrm{cm}$$

$$r_2 = 3 + 2 = 5\ \mathrm{cm}$$

$$r_3 = 5 + 2 = 7\ \mathrm{cm}$$

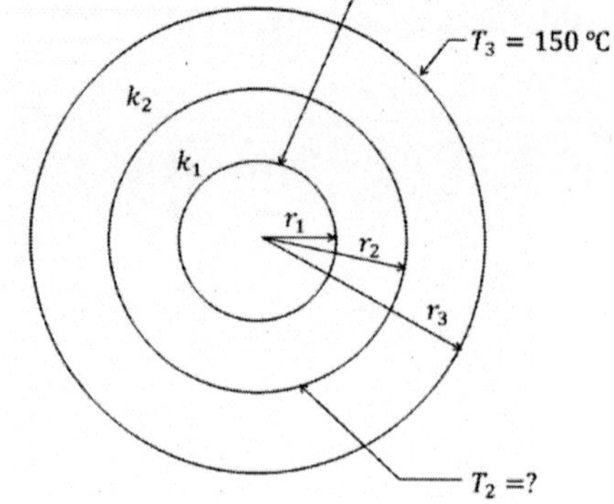

Figure 2.14: Radial Heat Flow Through a Hollow Cylinder

$$k_1 = 43\ W/m.{}^\circ\mathrm{C}$$

$$k_2 = 4.15\ W/m.{}^\circ\mathrm{C}$$

$$T_1 = 800\ ^\circ\mathrm{C}$$

$$T_3 = 150\ ^\circ\mathrm{C}$$

Calculate, $^q/_L$ and T_2

where, L is the length of the cylinder.

$$q/_L = \frac{2\pi(T_1 - T_3)}{\ln(^{r_2}/_{r_1})/k_1 + \ln(^{r_3}/_{r_2})/k_2}$$

$$= \frac{2\pi(800 - 150)}{\ln(^5/_3)/43 + \ln(^7/_5)/4.15}$$

$$= \frac{2\pi \times 650}{0.0118 + 0.0810} = \frac{2\pi \times 650}{0.0928}$$

$$= 43934.90\ W/m$$

This heat flow is used to calculate the interface temperature,

$$q/_L = \frac{2\pi(T_2 - T_3)}{\ln(^{r_3}/_{r_2})/k_2}$$

$$43934.90 = \frac{2\pi(T_2 - 150)}{\ln(^7/_5)/4.15}$$

$$43934.90 = \frac{2\pi(T_2 - 150)}{0.0810}$$

$$T_2 = \frac{43934.90 \times 0.0810}{2\pi} + 150$$

$$= \mathbf{716.93}\ °\mathrm{C}$$

QUESTIONS

1. Differentiate between heat transfer by conduction, convection, and radiation.
2. State Fourier's law of heat conduction and derive the expression for one-dimensional heat flow through a circular hollow composite cylinder.
3. Define Reynolds number, Prandtl number and Nusselt number and give the relationship between them for a hot flat plate.
4. Two infinite black plate at 600°C and 200°C exchange heat by radiation. Calculate the heat transfer per unit area.

   (*Hint:* $q/A = \sigma(T_1^4 - T_2^4)$, $\sigma = 5.669 \times 10^{-8} \frac{W}{m^2} \cdot \frac{1}{K^4}$, $T_1 = 600 + 273 = 873$ K, $T_2 = 200 + 273 = 473$ K)

CHAPTER 3

Propeller Aerodynamics

3.1 INTRODUCTION

A propeller is a set of rotating blades, or wings, such that the direction of the resultant force is primarily forward. The propeller force provides thrust to push the airplane through the air. The blade is a twisted airfoil that converts rotating power of the engine into thrust, which propels the airplane. The propeller is rotated by a piston engine or a gas turbine. The pitch of the blade may be fixed or variable. The term pitch has been taken from the pitch of the screw, which is advance of the screw for one turn. For a propeller, rotating n times per second, the time for one rotation is $1/n$, during which propeller advances by V/n. This is defined as the pitch of the propeller. Here, V is flight speed. The tangent of the angle, β made by propeller advancing speed to propeller azimuthal speed is,

$$\tan\beta = \frac{\text{pitch}}{\pi D}$$

where, D = Propeller diameter

In the early development stage, the propellers were made by hand from wood. Then came the era of metals like aluminium, and in modern designs, composite materials are also used. Sometimes, between the prime mover and the propeller, a reduction gear unit is used. The gearbox is essentially required for turboprop, while for light aircraft this may not be necessary. A rotating airfoil, behind the aircraft, which pushes, is called a pusher, while the one which pulled from the front is called a tractor. Both these combinations were referred to as propeller or air screws. The efficiency of propeller has improved from the period of Wright brothers, i.e., early 20th century, from 82% to present age, more than 90%.

A propeller's performance changes when the blade speed nears transonic. As the relative speed at any section of a propeller is a vector sum of aircraft speed and the tangential speed due to rotation, a propeller blade tip will reach transonic speed well before the aircraft does. When the air flow over the tip of the blade reaches its critical speed, drag and torque resistance increases rapidly and shockwaves are formed creating a sharp increase in noise. Aircraft with conventional propellers usually do not fly faster than Mach 0.6. There have been propeller aircraft which attained up to the Mach 0.8 range, but the noise generated was tremendous.

A fan is a propeller with a large number of blades. A fan, therefore, produces a lot of thrust for a given diameter, but the closeness of blades means that each strongly affects the flow around the others. If the flow is supersonic, this interference can be beneficial. By placing the fan within a shaped duct, specific flow patterns can be created depending on flight speed and engine performance. As the air enters the duct, its speed is reduced while its pressure and temperature increases. A ducted fan or propeller is also advantageous at lower speeds, but the duct needs to be shaped in a different manner than one for higher speed flight. Noise is reduced by ducting. However, the duct adds weight, cost and drag.

The performance of a propeller can be studied by using either of the two basic theories. One is blade element theory and the other is momentum theory.

3.2 Blade Element Theory

According to blade element theory, each element of a blade is regarded as an airfoil moving in an appropriate manner. A propeller blade is twisted such that the chord line changes from being almost parallel to V_∞, the free stream velocity at the root, to almost perpendicular at the tip as shown in Figure 3.1

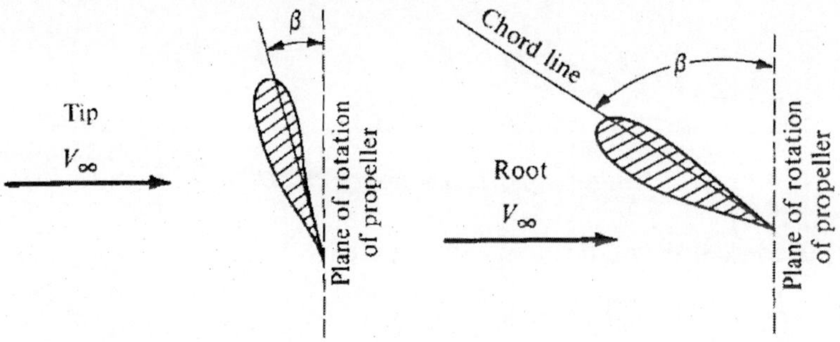

(a) Pitch at Tip of Propeller (b) Pitch at Root of Propeller

Figure 3.1

(*Source:* Anderson, Jr. John D., Introduction to Flight, Seventh Edition, McGraw-Hill 2012)

Pitch angle, β is the angle between the chord line and propeller's plane of rotation. β is a function of radius r, i.e., $\beta(r)$. Wind velocity is V_∞ and $r\omega$ is the speed of blade at section due to rotation of propeller. ω is the angular velocity in radians/sec. Then the relative wind speed as seen by the propeller section will be vector sum of V_∞ and $r\omega$. This is shown in Figure 3.2.

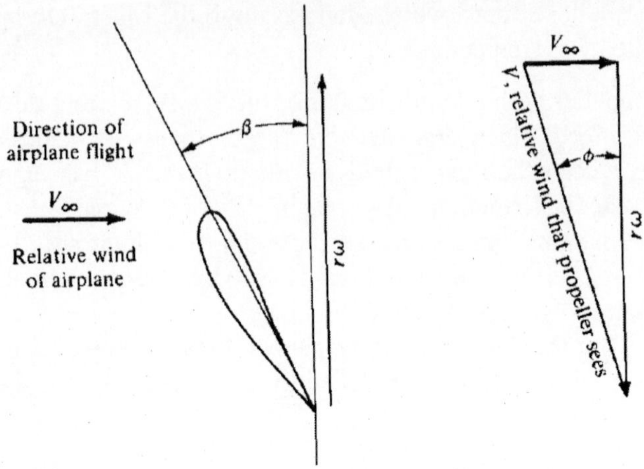

Figure 3.2: Velocity Diagram

Let, α = angle of attack,
L = Lift (perpenducular to V),
D = Drag (parallel to V)

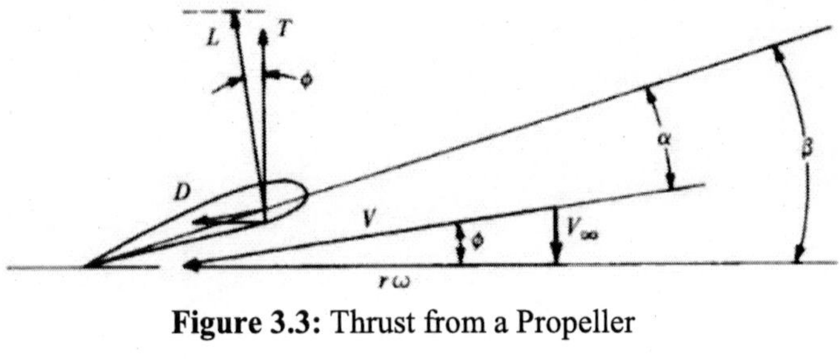

Figure 3.3: Thrust from a Propeller

When the chord line will be at an angle of attack α, Figure 3.3, the thrust T,

$$T = L\cos\varphi - D\sin\varphi$$

where, $\varphi = \beta - \alpha$

When the thrust over the entire blade length is added, will be the net thrust T_A which drives the airplane forward.

The propeller efficiency is defined as,

$$\eta = \frac{P_A}{P}$$

where, P = shaft brake power and P_A = Power available from the propeller

$$P_A = T_A . V_\infty$$

therefore,

$$\eta = \frac{T_A . V_\infty}{P}$$

The angular velocity,

$$\omega = 2\pi n$$

where, n = revolutions/ sec and, T_A = Thrust available from the engine

We define a dimensionless number, advance ratio, J

$$J = \frac{V_\infty}{nD}$$

It may be noted that η is always less than unity as some power delivered by the shaft to the propeller is always lost. The fastest moving part of the blade is the tip. It is near sonic. Then shock waves occur and boundary layer separation losses also take place.

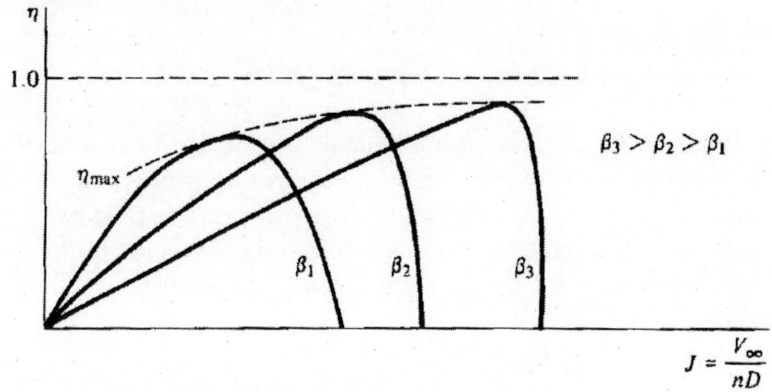

Figure 3.4: Plot Showing Relation between Propeller Efficiency and Advance Ratio for a Fixed β

Figure 3.4 gives the relation between J and η for a fixed β and when β is more, efficiency is better. Efficiency is always less than unity. When the propeller tip speed is near or more than sonic velocity, the efficiency will drop due to compressibility effects. The shock waves and boundary layer separation losses occur. This is why the propeller driven aircraft are not suitable for transonic and supersonic speeds.

When V_∞ is zero, $J = 0$ and efficiency of the propeller is zero. Also, when V_∞ is large, J is large, the propeller loses lift owing to small angle of attack and therefore, η = 0. This is shown in Figure 3.5. This occurs at the zero lift line. Further increase in V_∞ will result in negative lift. This also explains, why a blade is twisted, with large β at the root and small β at the tip. In the older days, up to 1930, fixed pitch propellers were made which were giving highest performance at the designed β and the propellers were mounted on the shaft.

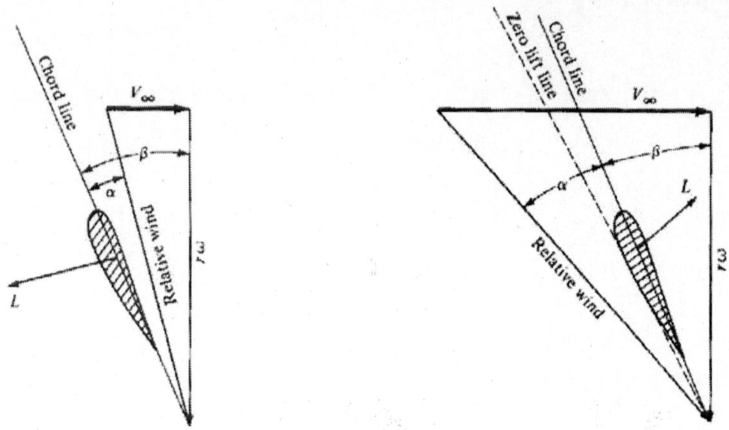

(a) Low V_∞ (b) High V_∞

Figure 3.5: Change of Propeller Efficiency with Advance Ratio

Propeller efficiency was highest at a particular value of J. At all other flight velocity the propeller efficiencies were lower. In the present era, variable pitch propellers are designed and the pilot can change the β value during flight. The propeller is mounted with a mechanism which rotate the entire blade about an axis along the length of the blade.

3.3 MOMENTUM THEORY

In this theory, the momentum and kinetic energy of the system is considered. The propeller is referred to as an actuator disc and slipstream rotation is ignored. The energy for rotation of propeller results in increase of velocity across the propeller (Figure 3.6).

It may be noted that the Bernoulli's equation is valid both at upstream and downstream of the propeller. The pressure of the fluid across the propeller increases by Δp.

Let $A =$ Area of the propeller

$V_0 =$ Upstream air velocity

$V_0 + aV_0 =$ velocity both upstream and downstream of propeller

$a, b =$ fraction by which velocity V_0 increases

$p_0 =$ pressure of air upstream and far off downstream

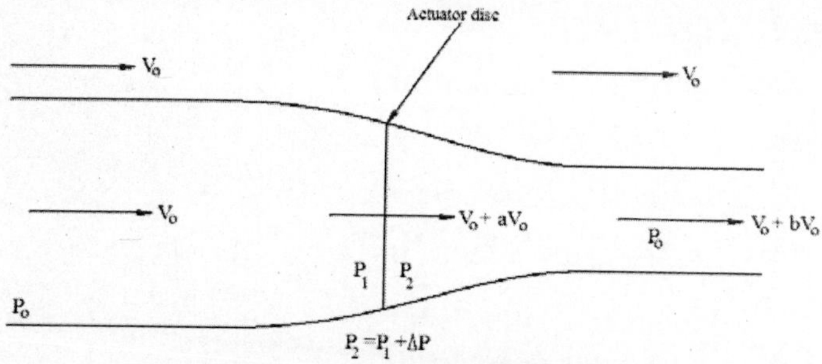

Figure 3.6: Actuator Disc Flow

(*Source:* Shevell Richard *S*, Fundamentals of Flight, Second Edition, Pearson, 1988)

$p_1 =$ pressure of air just upstream of propeller

$p_2 =$ pressure of air just downstream of propeller

$\rho =$ density of air

$p_T =$ total pressure

then, for upstream of propeller,

$$p_{T_1} = p_0 + \frac{1}{2}\rho V_0^2$$

$$p_{T_1} = p_1 + \frac{1}{2}\rho(V_0 + aV_0)^2$$

$$p_{T_2} = p_0 + \frac{1}{2}\rho(V_0 + bV_0)^2$$

$$p_{T_2} = (p_1 + \Delta p) + \frac{1}{2}(V_0 + aV_0)^2$$

Change in pressure at propeller disc,

$$\Delta p = p_{T_2} - p_{T_1}$$

$$\Delta p = \left[p_0 + \frac{1}{2}\rho(V_0 + bV_0)^2\right] - \left[p_0 + \frac{1}{2}\rho V_0^2\right]$$

$$= \frac{1}{2}\rho[V_0^2 + 2bV_0 + b^2V_0^2] - \frac{1}{2}\rho V_0^2$$

$$= \frac{1}{2}\rho[2bV_0 + b^2V_0^2]$$

$$= \rho V_0^2\left[b + \frac{b^2}{2}\right]$$

$$= \rho b V_0^2\left[1 + \frac{b}{2}\right]$$

Now thrust, T

$$T = \text{rate of change of momentum}$$
$$= \Delta p.A$$

also,

$$T = \rho A(V_0 + aV_0)[V_0 + bV_0 - V_0]$$

compare the two,

$$\Delta p = \rho(V_0 + aV_0).b\,V_0$$
$$= \rho V_0^2(1 + a).b$$

compare the two values of Δp

$$\rho b V_0^2\left[1 + \frac{b}{2}\right] = \rho V_0^2(1 + a)b$$

$$b + \frac{b^2}{2} = b + ab$$

$$\frac{b^2}{2} = ab$$

$$b = 2a$$

or,

$$a = b/2$$

This shows that 50% of velocity increase takes place just upstream of the propeller and 50% increase takes place far behind the propeller. Therefore, thrust,

$$T = \rho A(V_0 + aV_0).bV_0$$
$$T = \rho A(V_0 + aV_0).2aV_0$$
$$T = 2A\rho V_o^2(1 + a).a$$

Let m be mass of air flowing through propeller per unit time.

Then,

$$m = \rho A(V_0 + aV_0)$$

and increase in kinetic energy across the propeller,

$$\Delta K.E. = \frac{1}{2}m[(V_0 + bV_0)^2 - V_o^2]$$
$$= \frac{1}{2}\rho A(V_0 + aV_0)[(1 + b)^2 - 1].{V_0}^2$$

$$= \frac{1}{2}\rho A V_0 (1+a)(b^2+2b).{V_0}^2$$
$$= \frac{1}{2}\rho A V_0^3 (1+a)(b+2)b$$
$$= \frac{1}{2}\rho A V_0^3 (1+a)(2a+2).2a$$
$$= 2\rho A V_0^3 (1+a)^2.a$$
$$= T.V_0(1+a)$$

useful work done by the propeller,

$$\text{output} = T\,V_0$$

therefore, propeller efficiency,

$$\eta = \frac{T.V_0}{\Delta\,KE}$$
$$= \frac{T.V_0}{T.V_0(1+a)}$$
$$= \frac{1}{1+a}$$

This means that if a is higher, i.e., air flow across the propeller is more, the efficiency will be less. To get higher efficiency of the propeller, the propeller diameter should be large, to have higher mass flow rate of air with very small increase in velocity across the propeller. But higher the propeller dia, the more will be its weight. It may be noted that the momentum theory gives the ideal efficiency which can never be achieved in practice and it does not consider the blade cross section and design. A large propeller giving a small velocity increase to a large amount of air is more efficient than a small propeller.

The limitations of momentum theory are:

1. It ignores the friction drag of the blades,
2. the kinetic energy of rotation of the slip stream and,
3. the fact that the thrust is not uniformly distributed over the blade.

The analysis given below is in FPS (Foot, Pound, Second) system.

The thrust horsepower provided by the propeller is

$$hp = \frac{TV_0}{550}$$
$$= (bhp).\eta$$

where, T = Thrust, lb

V_o = Velocity, $\frac{ft}{s}$

bhp = engine brake horse power

The conversion factor,

$$1\ hp = 550\ ft.\frac{lb}{s}$$

hp = horse power, η = efficiency

In S.I. system, $1\ hp = 746\ W$

It may be noted that V_o and V_∞ are same.

We define another dimensionless number C_p, the power coefficient,

$$C_p = \frac{bhp \times 550}{\rho n^3 D^5}$$

and advance ratio, J has been defined as,

$$J = \frac{V_0}{nD}$$

where, D = propeller diameter,

n = revolutions per second

A plot between J and C_p is called a propeller chart.

A typical propeller chart is shown in Appendix - A for a four bladed propeller with a blade activity factor of 135. The activity factor is a parameter dependent on blade width weighed for the greater effectiveness of faster moving sections near the tip. For large, high speed turbo-props, a special form of propeller called prop fan gives better efficiency upto cruise Mach number of 0.8. The conventional propeller show efficiency decrease above Mach number 0.65–0.70. The high solidity propeller, i.e. many wide blades are used in unducted fans.

EXAMPLE

Example 3.1

A turboprop powered airplane flying at 304 knots at 20,000 ft on a standard day. (1 knot = 1 nautical miles/h = 0.5148 m/s). Each of the four 14 ft diameter propellers is driven by an engine delivering 1920 shaft power.

The propeller speed is 1050 rpm. Assume, that the propellers are four bladed with a blade activity factor of 135. Determine:

1. Propulsive efficiency
2. Thrust per propeller
3. Ideal efficiency from simple momentum theory

Solution

1. At 20,000 ft high, the density of air

$$\rho = 0.0012673 \text{ slug}/ft^3$$

$$[1lb = 1 \text{ slug} \times 32\ ft/s^2]$$

$$C_p = \frac{bhp \times 550}{\rho n^3 D^5}$$

$$= \frac{1920 \times 550}{0.0012673 \times (1050/60)^3 (14)^5}$$

$$= 0.289$$

Power is in $ft.lb/s, n$ is revolution per second and V in ft/s.

$$J = \frac{V_0}{nD} = \frac{304 \times 1.69}{(1050/60) \times 14} = 2.10$$

From propeller chart (Appendix - A),

$$\eta = 0.851$$

2.

$$T = \frac{550\ bhp.\eta}{V}$$

$$= \frac{550 \times 1920 \times 0.851}{304 \times 1.69} = 1751\ lb$$

3. Ideal efficiency

$$T = 2A\rho V_0^2(1 + a)a$$

$$1751 = 2\left[\frac{\pi}{4} \times 14^2\right] \times 0.0012673 \times (304 \times 1.69)^2(a + a^2)$$

$$= 102{,}985\ (a + a^2)$$

$$(a + a^2) - 0.017 = 0$$

This will give

$$a = \frac{-1 \pm 1.0334}{2}$$

$$= \frac{0.0334}{2} = 0.0167$$

Then,

$$\eta_{\text{ideal}} = \frac{1}{1+a}$$

$$= \frac{1}{1.0167}$$

$$= \mathbf{0.984}$$

3.4 PROPELLER SELECTION

For buying a propeller from off the shelf from a propeller manufacturer, we must calculate the specifications for a given application. For estimation of propeller diameter, the following empirical relations may be used. The propeller efficiency improves as the diameter of the propeller gets larger. For the same thrust, the larger propeller requires a smaller flow velocity increase across the propeller disk. The smaller the increase in flow velocity across the propulsive device, the higher the propulsive efficiency. However, there are two limitations, (i) the propeller tip must clear the ground, and (ii) the propeller tip speed should be less than the velocity of sound otherwise, the compressibility effects will occur. Two bladed propellers are common on general aviation aircraft. Three and four blade propellers are used in combat and commercial aircraft.

The empirical relations for a propeller of a diameter *D*, where *D* is in inches, are as follows:

Two-blade $D = 22\ (hp)^{1/4}$

Three-blade $D = 18\ (hp)^{1/4}$

where *hp* is horsepower of the engine.

For better understanding, an example is given below.

Example 3.2

Let us select a propeller for a 360 hp engine and choose a two-blade constant speed propeller.

$$D = 22(360)^{1/4}$$

$$= 95.83 \text{ in}$$

$$= 8 \text{ ft.}$$

Now, we will check the tip speed. Let the rpm be 2600. Then,

$$(V_{tip})_0 = \pi n D$$

$$= \pi \times \frac{2600}{60} \times 8$$

$$= 1089 \ \frac{ft}{s}$$

Assume, max forward speed, $V_\infty = 250$ mph

or,

$$V_\infty = 366.7 \ \frac{ft}{s}$$

Then the tip velocity will be

$$V_{tip} = \sqrt{\left(\left(V_{tip}\right)_0^{\ 2} + V_\infty^{\ 2}\right)}$$

$$= \sqrt{(1089)^2 + (366.7)^2}$$

$$= 1149 \ \frac{ft}{s}$$

The speed of sound at standard sea level is 1117 ft/s and, therefore, the tip speed is more than the velocity of sound, which is not desirable.

Now, select a 3-blade propeller,

$$D = (18 \text{ hp})^{1/4}$$

$$= 18(360)^{\frac{1}{4}} = 78.4 \text{ in}$$

$$= 6.53 \text{ ft}$$

then the tip speed will be,

$$\left(V_{tip}\right)_0 = \frac{\pi n D}{60} = \pi \times \frac{2600}{60} \times 6.53$$

$$= 889 \text{ ft/s}$$

therefore,

$$V_{tip} = \sqrt{\left(\left(V_{tip}\right)_0^{\ 2} + V_\infty^{\ 2}\right)}$$

$$= \sqrt{(889^2 + 366.7^2)}$$

$$= 962 \text{ ft/s}$$

which is acceptable. Therefore, a 3-blade propeller of 6.53 ft. diameter is selected/ recommended.

3.5 HELICOPTER ROTOR

The helicopter rotor is the combination of several rotary wings (rotor blades) and a control system that generates the aerodynamic lift force which supports the weight of the helicopter, and the thrust to counter aerodynamic drag in forward flight. A helicopter is one of the examples of rotary wing aircraft. The name is derived from the Greek word helix, meaning spiral and wing. Most helicopters maintain a constant rotor speed during flight leaving the angle of attack of the blades as the sole means for adjusting thrust from the rotor. Helicopters have a single main rotor but require a separate rotor to average torque. This is accomplished through a variable pitch antitorque rotor or tail rotor. The materials of construction of blade are aluminium, composite structure and steel or titanium with abrasion shields along the leading edge. In a helicopter, the rotor disk is oriented such that we call it force 'lift' rather than 'thrust'. On a helicopter, the force of a tail rotor would be best described as thrust. The solidity ratio is the ratio of the total rotor blade area, which is the combined area of all the main rotor blades, to the total rotor disk area.

SUMMARY

A propeller is a set of rotating blades, or wings, such that the resultant force is primarily forward. The blade is a twisted airfoil that converts rotating power of the engine into thrust, which propels the airplane. The propeller is rotated by a piston engine or gas turbine. The performance of a propeller can be evaluated by using either of the two basic theories, one is blade element theory another is momentum theory. In blade element theory, each element of the blade is referred to as an airfoil moving in an appropriate manner. A propeller blade is twisted such that the chord line changes from being almost parallel to V_∞, the free stream velocity at the root, to almost perpendicular at the tip.

In momentum theory, the momentum and kinetic energy of the system are considered. The propeller is referred to as an actuator disk. The energy for rotation of propeller results in increase of velocity across the propeller. To get higher efficiency of the propeller, the propeller diameter should be large, to have higher mass flow rate of air with very small increase in velocity across the propeller. But higher the propeller diameter more will be its weight. A large propeller giving a small velocity increase to a large amount of air is more efficient than a small propeller.

Two dimensionless numbers are defined, the power coefficient, C_p

$$C_p = \frac{bhp \times 550}{\rho n^3 D^5}$$

and, advance ratio, J

$$J = \frac{V_0}{nD} = \frac{V_\infty}{nD}$$

The empirical relations for a propeller of diameter D, where D is in inches, are as follows,

Two-blade, $D = 22(hp)^{1/4}$

Three-blade, $D = 18(hp)^{1/4}$

A brief discussion on helicopter rotor is also given.

QUESTIONS

1. Define pitch of propeller and discuss the materials of construction.
2. Explain the blade element theory and derive the expressions for thrust generated and power absorbed by the propeller.
3. State the actuator theory of propellers. Derive the expressions for the thrust generated and the power required.
4. Define advance ratio, J and power coefficient C_p for a propeller. Draw a typical propeller chart and explain its use.

CHAPTER 4

Combustion

4.1 INTRODUCTION

Combustion is a fast exothermic gas phase reaction, generating heat or both heat and light, where fuel and oxidizer are the reactants. A flame is a combustion reaction which can propagate subsonically through space and it propagates with respect to unburned gases. In case of internal combustion engines, the flame is confined to a small zone in the combustion chamber. This reaction zone is called flame front. When fuel and oxidizer are premixed before ignition, the flame is called premixed flame and when fuel and oxidizer mix in the same region, the flame is called a diffusion flame as mixing takes place due to diffusion. Another way of classifying flames is laminar or turbulent. In laminar or streamlined flow, the mixing and, therefore, transport are by molecular process. Laminar flow occurs at low Reynolds number. The Reynolds number is the ratio of inertial forces to viscous forces. In turbulent flows, mixing and transport are enhanced by macroscopic relative motion of the eddies. When the flame structure and motion do not change with time, it is called steady flame and otherwise unsteady flame.

An undesirable combustion phenomenon is the spontaneous ignition of a substantial mass of fuel air mixture ahead of the flame, called knocking. This results in generation of high pressure and damage to the engine. In case of spark ignition engine, the knocking takes place in the later part of combustion process and for compression ignition engine, it takes place in the early part of combustion due to ignition delay and flooding of fuel in the combustion chamber.

4.2 FUELS

A fuel may be solid, gaseous or in a liquid form. Solid fuels do not find much application in internal combustion engines due to the problem of handling and disposing off ash after combustion. Handling, storage and feeding of solid fuel is also cumbersome. Gaseous fuels are ideal for internal combustion engines. They mix more homogeneously with air. However, its storage and handling problems restrict the use. Most of the internal combustion engines use liquid fuels which are basically derivatives of liquid petroleum. Petroleum is a mixture of many hydrocarbons with different molecular structure. It also contains oxygen, nitrogen, sulphur and impurities. The carbon and hydrogen combine in different proportions and molecular structure to form a hydrocarbon. The hydrogen carbon ratio is important for defining the heating value or energy content of the fuel. More hydrogen means higher energy content.

Both gaseous and liquid hydrocarbon fuels are a mixture of many different hydrocarbons. Gasoline is a mixture of some 40 hydrocarbons. For combustion calculation, gasoline fuel is considered as octane or isoctane, C_8H_{18} and diesel fuel as do-decane, $C_{12}H_{26}$. Two hydrocarbons with same number of hydrogen and carbon atoms and different structures are called isomers. Table 4.1 gives the basic hydrocarbon families.

Table 4.1: Basic Hydrocarbon Families

Family of Hydrocarbons	***General Formula***	***Molecular Structure***	***Saturated/Unsaturated***	***Stability***
Paraffin	C_nH_{2n+2}	Chain	Saturated	Stable
Olefin	C_nH_{2n}	Chain	Unsaturated	Unstable
Naphthene	C_nH_{2n}	Ring	Saturated	Stable
Aromatic	C_nH_{2n-6}	Ring	Highly unsaturated	Most unstable

Source: Ganesan V., Internal Combustion Engines, Mc Graw Hill, Fourth Edition, 2013.

Some other organic compounds used as fuel include dimethyl ether and diethyl ether. The main properties which need consideration for selection of fuel include volatility, vapor lock, antiknock quality and energy content or heating value. Volatility is the main characteristic property of gasoline which finds its suitability in a spark ignition engine. For compression ignition engine, the fuel should be sufficiently volatile in the operating range of temperature for good mixing and combustion. High rate of vaporization of gasoline fuel may upset the carburetor metering or even stop the flow due to vapour lock in the fuel passage. High boiling temperature hydrocarbons are better fuels. Abnormal combustion i.e. knocking is not desirable in any internal combustion engine. Therefore, antiknock property is important. For gasoline, it is defined in terms of octane number. The number is estimated with a mixture of two reference fuels, isooctane C_8H_{18} and normal heptane C_7H_{16}. Isooctane has been assigned an arbitrary rating of 100 octane number and normal heptane, which has very poor antiknock quality is given a rating of zero octane number. The octane number is defined as percentage by volume of isooctane in a mixture of iso-octane and normal heptane, which exactly matches the knocking intensity in a standard engine under a set of standard operating conditions.

For diesel fuel, cetane number is a measure of fuel's ability to autoignite. Low cetane number signifies relatively long ignition delay and higher cetane number implies relatively short ignition delay. The test is carried out in a standard engine to measure ignition delay. The cetane number of n-cetane (C_6H_{34}) is defined as 100, and that of heptamethyl nonane (also C_6H_{34}) as 15. The cetane number is defined as sum of vol% n-cetane and 0.15 vol% heptamethyl nonane. The cetane number ranges from 0–100. A typical fuel has cetane number in the range 30–65. For continuous combustion systems such as jet engines, ignition quality is less important than in reciprocating engines. In aircraft jet engines, re-lighting of engines during inflight is important. The energy density or lower heating value of some common fuels is given in Table 4.2.

Table 4.2: Lower Heating Value of Some Common Fuels

Fuel	*Formula*	*Lower Heating Value, kJ/kg*
Gasoline	$C_nH_{1.87n}$	44,000
Light diesel	$C_nH_{1.80n}$	42,500
Isoctane	C_8H_{18}	44,300
Methanol	CH_3OH	20,000
Ethanol	C_2H_5OH	26,964
Hydrogen	H_2	116,000

Source: Ganesan V., Internal Combustion Engines, Mc Graw Hill, Fourth Edition, 2013.

Hydrocarbon fuels may contain some sulphur, which is not desirable. During combustion with oxygen, sulphur dioxide is formed and in presence of water at low temperature, may form sulphuric acid. For ease of flow in pipelines, density and viscosity of fuel also need consideration.

4.3 GAS TURBINE FUELS

In modern turbojet engines, the fuel flow rate is high and they operate with wide temperature and pressure variations. A typical jet fuel contains 16 percent hydrogen and 84 percent carbon, produced from crude petroleum oil. Some impurities like sulphur, nitrogen, water and sediment are also present. JP-4 is the most commonly used fuel. JP-5 is a denser, less volatile fuel than JP-4 and can be safely stored in tanks. The common commercial aircraft fuels are Jet A and Jet A1. Both are alike, but freezing point of Jet A1 is lower than Jet A. The heating value of most of the jet fuels is 42,800 kJ/kg. the heating value of hydrogen is 116,000 kJ/kg and is suitable for high Mach number flight like Scramjet.

Table 4.3: Data for JP Fuels

Fuel	*Formula*	*Flash Point, °C*	*Boiling Point, °C*
JP-4	C_8H_{16}	–	207
JP-5	$C_{12}H_{23}$	60	247
JP-8 (Jet A1)	$C_{10}H_{21}$	38	252

JP 4, JP 5 and JP 8 are also called aviation kerosene, and are a blend of kerosene and gasolene.
Source: Bose, Tarit, Airbreathing Propulsion—An Introduction, Springer, 2012.

4.4 PROPELLANT

When the fuel and oxidizer are together, it is called a propellant. They are classified as liquid propellant and solid propellant. The liquid propellent is carried on board and is injected under pressure into the combustion chamber. The injection pressure may be as high as 440 atm. Due to high pressure, the storage tanks are quite heavy. Another way of feeding is pump-fed system. An example of liquid propellant is liquid hydrogen and liquid oxygen. They are designated as cryogenic propellants as they are stored at very low temperatures, hydrogen at 20 K and oxygen at 135 K. The combustion process is started with an igniter and then the burning is self sustaining.

A typical solid propellant consists of atomized aluminium powder (16 percent) as a fuel and ammonium perchlorate (69.93 percent) as an oxidizer. The remainder is iron-oxide powder (0.7 percent) as a catalyst and polybutadiene acrylic acid acrylonitride (about 14 percent) as a rubber based binder. The binder also burns as a fuel.

4.5 COMBUSTION PROCESS

The reaction of fuel and oxidizer is combustion and during the process, the mass of each element remains the same. The simplest reaction is between carbon and oxygen.

$$C + O_2 \rightarrow CO_2$$

Here, 1 kmol of carbon reacts with 1 $kmol$ of oxygen to form 1 kmol of carbon dioxide. On mass basis, 12 kg of carbon reacts with 32 kg of oxygen to form 44 kg of carbon dioxide. Now, examine the combustion of methane,

$$\mathrm{CH_4 + 2O_2 \rightarrow CO_2 + 2H_2O}$$

The products of combustion are carbon dioxide and water. The water may be in vapour, liquid or solid phase, depending on the temperature and pressure of the products of combustion.

In most of the combustion processes, oxygen is supplied as air. The composition of air on molal or volume (both are same) basis is approximately 21% oxygen, 78% nitrogen and 1% argon. It is assumed that nitrogen and argon do not take part in the chemical reaction. Dissociation of gases takes place at high temperature. At high temperatures, produced in internal combustion engines, reaction between nitrogen and oxygen takes place and nitrogen oxides are formed which creates air pollution problems.

For combustion calculations, air will be considered as a mixture of 21% oxygen and 79% nitrogen. Therefore, for every mole of oxygen, 79/21 = 3.76 moles of nitrogen is present. The combustion reaction for methane with air,

$$CH_4 + 2O_2 + (2 \times 3.76)N_2 \rightarrow CO_2 + 2H_2O + 7.52N_2$$

This is complete combustion equation and products contain no oxygen. This also gives the theoretical or stoichiometric requirement of air.

For complete combustion of Octane, C_8H_{18} with air,

$$C_8H_{18} + 12.5O_2 + (12.5 \times 3.76)N_2 \rightarrow 8CO_2 + 9H_2O + 47.0N_2$$

The air-fuel (A/F) ratio on mole basis

$$\frac{A}{F} = \frac{(12.5) + (12.5 \times 3.76)}{1}$$

$$= 59.5 \text{ kmol air/kmol fuel}$$

The theoretical air-fuel ratio on mass basis,

$$\frac{A}{F} = \frac{(12.5 \times 32) + (12.5 \times 3.76) \times 28}{(8 \times 12) + (18 \times 1)}$$

$$= 15.0 \text{ kg air/kg fuel}$$

When octane fuel is burned with 100% excess air,

$$C_8H_{18} + (12.5 \times 2)O_2 + (12.5 \times 3.76 \times 2)N_2$$
$$\rightarrow 8CO_2 + 9H_2O + 12.5O_2 + 94.0N_2$$

Total $kmol$ of product

$$= 8 + 9 + 12.5 + 94.0 = 123.5$$

Therefore, molal analysis of product:

$$CO_2 = 8/123.5 = 6.47\%$$

$$H_2O = 9/123.5 = 7.29\%$$

$$O_2 = 12.5/123.5 = 10.12\%$$

$$N_2 = 94/123.5 = 76.12\%$$

$$\text{Total} = 100.00$$

The equivalence ratio, ϕ is defined as,

$$\phi = \frac{(F/A)_{\text{actual}}}{(F/A)_{\text{stoichiometric}}}$$

The inverse of equivalence ratio is defined as λ,

$$\lambda = \frac{(A/F)_{\text{actual}}}{(A/F)_{\text{stoichiometric}}}$$

For fuel-lean mixture, $\phi < 1$
For stoichiometric mixture, $\phi = 1$
For fuel rich mixture, $\phi > 1$

The enthalpy of formation of a compound is defined as the enthalpy increase when one mole of the compound is formed from its elements, with each substance in the thermodynamic standard state at the given temperature, the standard state at one atmospheric pressure and temperature under consideration. The heating value or calorific value of a fuel is the magnitude of heat of reaction at constant pressure or at constant volume at a standard temperature (usually 25°C) for complete combustion of unit mass of fuel. Complete combustion means that all carbon is converted to CO_2 and all hydrogen is converted to H_2O and any sulphur present is converted to SO_2.

The term higher heating value or gross heating value is used when H_2O formed is condensed to liquid phase and the term lower heating value or net heating value is used when H_2O formed is in vapour state.

When combustion process takes place adiabatically and no work or change in kinetic or potential energy is involved, the temperature of the products is designated as adiabatic flame temperature. For a given fuel and predetermined pressure and temperature, the maximum adiabatic flame temperature is achieved with stoichiometric mixture. In gas turbine application, the adiabatic flame temperature is controlled by excess air.

SUMMARY

Combustion is a fast exothermic gas phase reaction, generating heat or both heat and light, where fuel and oxidizer are the reactants. A fuel may be solid, gaseous or in liquid form. The liquid fuels are basically derivatives of liquid petroleum. The hydrogen carbon ratio in fuel is important for defining the heating value or energy content of the fuel. More hydrogen means higher energy content. Gasoline fuel is considered as octane or iso-octane, C_8H_{18}. The main properties which need consideration for selection of fuel include volatility, vapour lock, antiknock quality and energy content or heating value. Hydrocarbon fuels may contain some sulphur, which is not desirable. A typical jet fuel contains 16% hydrogen and 84% carbon. The heating value of most of the jet fuels is 42,800 kJ/kg.

When fuel and oxidizer are together, it is called a propellant. The reaction of fuel and oxidizer is combustion and during the process the mass of each element remains the same. For complete combustion of octane, C_8H_{18} with air,

$$C_8H_{18} + 12.5O_2 + (12.5 \times 3.76)N_2 \rightarrow 8CO_2 + 9H_2O + 47N_2$$

The air fuel ratio on mass basis is 15.0 kg air/kg fuel.

The equivalent ratio, φ is defined as,

$$\varphi = \frac{(F/A)_{\text{actual}}}{(F/A)_{\text{Stoichiometric}}}$$

The value of $\varphi < 1$ means fuel lean mixture and $\varphi > 1$ for fuel rich mixture.

QUESTIONS

1. Discuss the classification of liquid fuels.
2. Write short notes on:
   (a) Solid and liquid propellants
   (b) Aviation gasoline
   (c) Vapor lock
   (d) Gas turbine fuels
   (e) Propellant
   (f) Stoichiometric fuel-air mixture ratio

CHAPTER 5

Internal Combustion Engines

5.1 INTRODUCTION

The years 1876, when Otto developed the spark ignition engine and 1892, when Diesel invented the compression ignition engine, are considered as the beginning of internal combustion engines. In these engines, the chemical energy of fuel is converted to thermal energy, which then produces mechanical power. The fuel is burned inside the engine. The hot products of combustion is the working fluid to produce mechanical energy. The engines have found wide application in transportation sector (land, sea and air) and power generation.

5.2 HISTORICAL

Heat engines were developed some two and half centuries ago. Water raised to steam was the working fluid. One of the main features of the early internal combustion (I.C.) engines was burning of fuel at atmospheric pressure. There was no compression of air before combustion of fuel. Combustion of fuel increased the pressure of the working fluid, which was expanded in a piston engine to produce power. The efficiency of such engines was about 5%. Nicolaus A. Otto (1832–1891) was the first to propose engine cycle with four strokes: intake, compression before ignition, expansion or power stroke and an exhaust stroke. In power stroke the work was delivered to the crankshaft. This has resulted in substantial reduction in weight and volume. Otto is regarded as the father of the modern internal combustion engines. Two stroke engines were then developed, where the exhaust and intake processes occur during the end of power stroke and beginning of compression stroke. The compression ratio was limited to 4 to avoid the knocking problem.

Rudolf Diesel (1858–1913), the German engineer, developed the compression engines, where the air was compressed, and the liquid fuel was injected in hot compressed air and combustion was initiated. This has resulted in greater expansion without detonation or knock increase and improved engine efficiency.

Fuels had major impact on engine development. For gasoline and lighter fractions of crude oil, various types of carburetors were developed to vaporize the fuel and mix with air. The high volatile fuel made starting easy and gave better cold weather performance. Compression ratios were low (4 or less) to avoid knocking. Kerosene was a good fuel for internal combustion engines, but it was not available because of its use for domestic purpose to produce heat and light. Over the years, fuels with better antiknock property have been introduced which has resulted in use of high compression ratios and produce higher power with better efficiency.

An I.C. engine is a reciprocating engine, where the piston moves back and forth in cylinder and transmits power through a connecting rod and crank mechanism to drive the shaft. Figure 5.1 gives the basic geometry of the engine. The piston moves between the top dead center (TDC) and the bottom dead center (BDC). The clearance volume,V_c and the swept volume, V_s are also indicated. The bore is the diameter of the piston and stroke is the displacement length of the piston. The ratio of maximum volume to minimum volume is called the compression ratio $((V_s + V_c)/V_c)$. The compression ratio for compression ignition (C.I.) engine is much higher than spark ignition (S.I.) engine. In a four-stroke engine, two revolutions are required to complete the cycle which produces one power stroke. To obtain a higher power stroke from a given engine cycle two-stroke engines were developed.

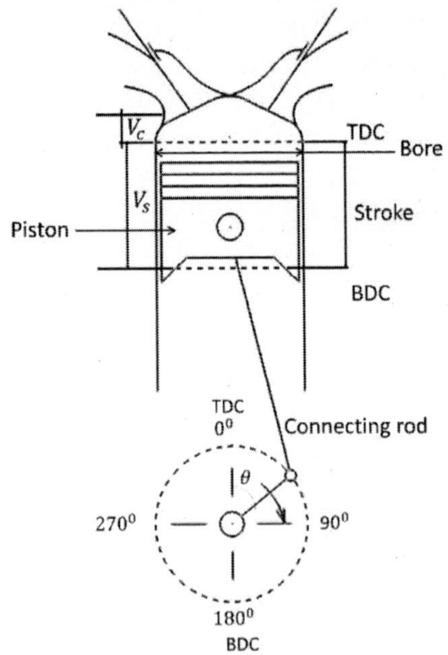

Figure 5.1: Sketch of Reciprocating I.C. Engine Showing Cylinder Clearance Volume, V_c and Swept Volume, V_s

(*Source:* Heywood, John B., Internal Combustion Engine Fundamentals, Mc Graw Hill Education (India) Private Limited, Chennai, 2017)

Two stroke cycle is applicable to both S.I and C.I engines. In two stroke engine there is one power stroke in every rotation. In this engine, the compression stroke, which starts by closing the inlet and exhaust ports compresses the working fluid and draws fresh charge into the crank case by opening the inlet port. As the piston approaches the TDC, combustion is initiated. A power or expansion stroke is similar to the four-stroke cycle until the piston approaches BDC. Most of the burned gases go out of the cylinder. When the inlet ports open, the fresh charge flows into the cylinder. It is difficult to fill completely the displaced volume and some of the fresh mixture flows out of the cylinder during the scavenging process.

5.3 SPARK IGNITION ENGINE

In S.I. engines the premixed fuel and air using a carburetor, or a fuel injection system is introduced into the engine cylinder. The stoichiometric air-fuel ratio for octane fuel is about 15 and is maintained to ensure homogeneous and stable combustion. The mixing of fuel is carried out in the carburetor. Fuel injection into the intake manifold or inlet port is an alternative to a carburetor. Figure 5.2 gives the plot of crank position against pressure rise in the cylinder. Location of spark plug is also given.

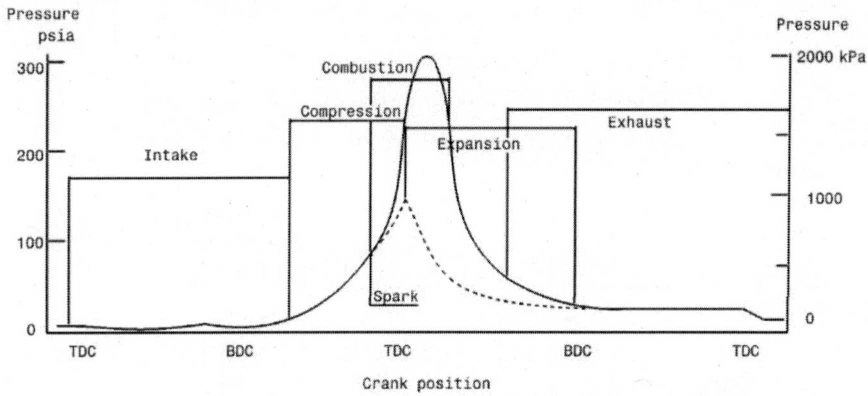

Figure 5.2: Plot of Crank Position with Pressure Rise for a Four-Stroke S.I. Engine

5.4 COMPRESSION IGNITION ENGINE

In C.I. engine, only air is inducted into the cylinder and compressed. The fuel is a light diesel oil and is injected directly into the engine cylinder in compressed air just before the combustion process is required to start. Load control is carried out by regulating the quantity of fuel injected in each cycle. The quantity of air remains unchanged. The liquid fuel evaporates and fuel vapor mixes with air. The air temperature and pressure are above the fuel ignition point. After a short delay, spontaneous ignition initiates the combustion process and the cylinder pressure increases. As the expansion proceeds, mixing between fuel-air and burning continues. The exhaust process is similar to that of four-stroke S.I. engine. At the end of exhaust stroke, cycle starts again.

5.4.1 Comparison of Spark Ignition and Compression Ignition Engines

The salient features of the S.I. engine are:

1. It works on Otto cycle i.e., constant volume heat addition.
2. Compression ratio is low, 6 to 10 depending on antiknock quality of fuel.
3. A highly volatile fuel like gasoline is used. The self-ignition temperature of fuel should be high.
4. Fuel air is pre-mixed before injection into the engine. A carburetor and ignition system are required. Modern engines use gasoline injection. A spark plug is necessary to ignite the fuel.
5. Throttle is used to control the fuel air mixture to control the load.
6. Due to lightweight and homogeneous combustion, the engine speed is high.
7. The thermal efficiency is low because of the low compression ratio.

The salient features of the C.I. engine are:

1. It works on Diesel cycle i.e., constant pressure heat addition.
2. The compression ratio is high, 16 to 20 and is limited on the basis of weight of the engine.
3. Diesel oil, a non-volatile fuel, is used and self-ignition temperature of the fuel should be low.
4. Air is compressed and then fuel is injected directly into the combustion chamber at the end of compression stroke. A fuel pump and injector are required. Self-ignition occurs because of high temperature of air after compression. No ignition system or spark plug is required.
5. The quantity of fuel is regulated to control the load.
6. Due to heavy weight and heterogeneous combustion, the engine speed is low.
7. Due to high compression ratio, the thermal efficiency is higher.

5.4.2 Comparison of Four-Stroke and Two-Stroke Engines

The salient features of the four-stroke engine are:

1. There are four strokes in a cycle and the cycle is completed in two revolutions of the crankshaft. Thus, there is one power stroke in two revolutions. Due to this, the turning moment is not uniform, and a heavier flywheel is required.
2. Cooling and lubrication requirements are low. Wear and tear are less.
3. Valves and valve actuating systems are used.
4. Due to higher weight, the initial cost of engine is high.
5. The volumetric efficiency is high.
6. The thermal efficiency is higher and part load efficiency is better.
7. Used where efficiency is important. Widely used in land transport and airplanes.

The salient features of the two-stroke engine are:

1. The thermodynamic cycle is completed in two strokes that is in one revolution of the crankshaft. Thus, there is one power stroke in every revolution of the crankshaft. The turning moment is more uniform, and the lighter flywheel will serve the purpose.
2. Cooling and lubrication requirements are higher, and wear and tear are also more.
3. There are no valves and ports are fitted.
4. Due to lighter weight, the engine cost is low.
5. The volumetric efficiency is low due to lesser time for mixture intake.
6. The thermal efficiency is lower and part load efficiency is poor.
7. Used where low cost, compactness and lightweight are important. An example is a two-wheeler.

5.5 AIR STANDARD CYCLES

An I.C. engine is an open system i.e. The working fluid enters the system at one set of conditions and leaves at another. The sequence of processes in the engine are intake, compression, combustion, expansion, and exhaust. For analysis purpose, a closed system is assumed. The following assumptions are made to analyze the air standard cycle.

1. The working fluid is a perfect gas and follows the relation, $p = \rho RT$.
2. The mass of fluid does not change during the cycle.
3. All the processes are reversible adiabatic i.e., isentropic.

4. Heat is supplied from a constant high temperature source and not from chemical reactions as in the real system.
5. Some heat is rejected to a constant low temperature sink.
6. No heat loss between system and the surroundings.
7. Specific heat of the working fluid remains constant during the cycle.
8. The working fluid is assumed to be air.

With these assumptions, the cycle efficiency will be the maximum achievable and in the real cycle, the efficiency will be much lower.

5.5.1 The Otto Cycle

In this cycle the heat addition process is at constant volume. The p-V and T-s diagrams are shown in Figure 5.3 (a) and (b).

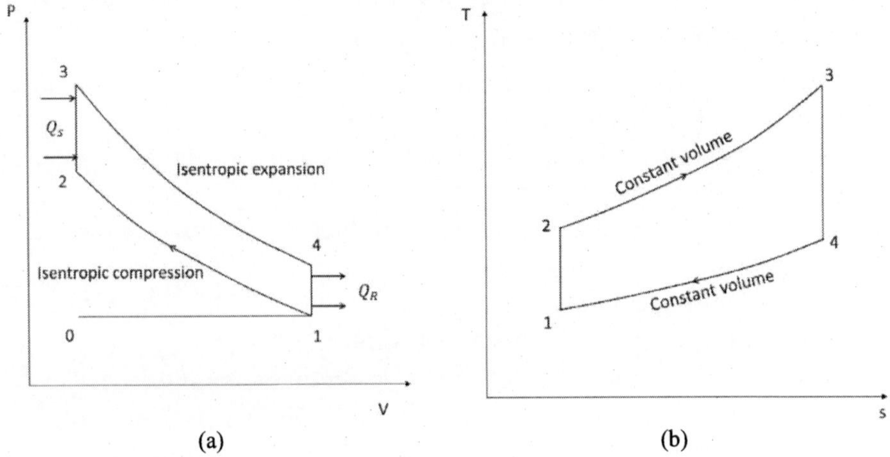

Figure 5.3: The Otto Cycle (a) p-V Diagram (b) T-s Diagram

(*Source:* Ganesan V., Internal Combustion Engines, McGraw Hill Education (India) Private Limited, New Delhi, Fourth Edition, 2013)

The processes:

0–1 Intake

1–2 Compression-Isentropic

2–3 Heat addition at constant volume, Q_S

3–4 Expansion-Isentropic-Power stroke

4–1 Heat rejection at constant volume, Q_R

1–0 Exhaust

Thermal efficiency,

$$\eta_{otto} = \frac{Q_S - Q_R}{Q_S}$$

$$= \frac{mC_v(T_3 - T_2) - mC_v(T_4 - T_1)}{mC_v(T_3 - T_2)}$$

$$= 1 - \frac{T_4 - T_1}{T_3 - T_2}$$

For isentropic processes 1–2 and 3–4,

$$\frac{T_2}{T_1} = \left(\frac{V_1}{V_2}\right)^{(\gamma-1)} = r^{(\gamma-1)}$$

and

$$\frac{T_3}{T_4} = \left(\frac{V_4}{V_3}\right)^{(\gamma-1)} = r^{(\gamma-1)}$$

$$\left(as, \frac{V_1}{V_2} = \frac{V_4}{V_3} = r\right)$$

where, $r =$ compression ratio

$$\frac{T_2}{T_1} = \frac{T_3}{T_4}$$

$$\frac{T_3}{T_2} = \frac{T_4}{T_1}$$

therefore,

$$\eta_{otto} = 1 - \frac{T_1\left(\frac{T_4}{T_1} - 1\right)}{T_2\left(\frac{T_3}{T_2} - 1\right)}$$

$$= 1 - \frac{T_1}{T_2}$$

$$= 1 - \frac{1}{r^{(\gamma-1)}}$$

This shows that the efficiency of the cycle is a function of compression ratio r and specific heat ratio, γ only. For various values of γ, the variation of efficiency with compression ratio r is given in the figure 5.4. For working fluid with higher values of γ, efficiency will be better.

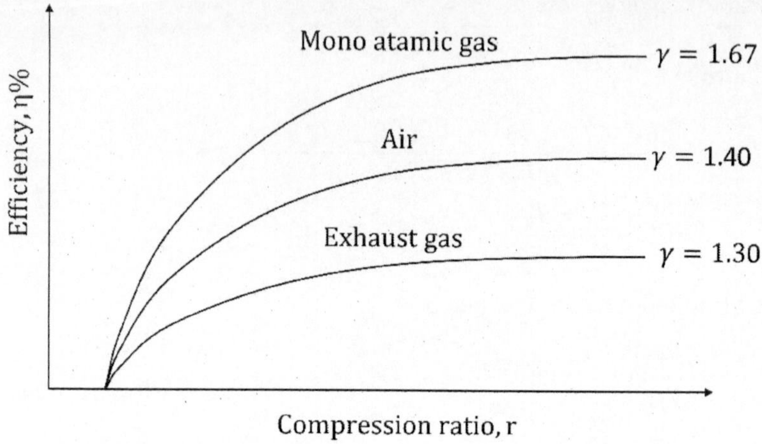

Figure 5.4: Effect of r and γ on Efficiency for Otto Cycle

Work output

Net work output for an Otto cycle:

$$W_N = W_E - W_C$$

where, W_C is compressor work, W_E is engine work and W_N is the net work output.

$$W_N = \frac{p_3V_3 - p_4V_4}{\gamma - 1} - \frac{p_2V_2 - p_1V_1}{\gamma - 1}$$

also,

$$\frac{p_2}{p_1} = \frac{p_3}{p_4}$$

$$= r^\gamma$$

$$(as, p_1V_1^\gamma = p_2V_2^\gamma)$$

$$(as, p_3V_3^\gamma = p_4V_4^\gamma)$$

therefore,

$$\frac{p_3}{p_2} = \frac{p_4}{p_1} = r_p\ (say)$$

also,

$$V_1 = rV_2$$

$$V_4 = rV_3$$

$$W_N = \frac{p_1V_1}{\gamma - 1}\left(\frac{p_3V_3}{p_1V_1} - \frac{p_4V_4}{p_1V_1} - \frac{p_2V_2}{p_1V_1} + 1\right)$$

$$= \frac{p_1V_1}{\gamma - 1}\left(\frac{r_p.r^\gamma}{r} - r_p - \frac{r^r}{r} + 1\right)$$

$$= \frac{p_1V_1}{\gamma - 1}(r_p - 1)(r^{\gamma-1} - 1)$$

Mean effective pressure, p_m

$$p_m = \frac{\text{work output}}{\text{swept volume}}$$

$$\text{swept volume} = V_1 - V_2$$

$$= V_2(r - 1)$$

$$p_m = \frac{p_1 r(r_p - 1)\left(r^{(\gamma-1)} - 1\right)}{(\gamma - 1)(r - 1)}$$

This shows that the net work output is directly proportional to pressure ratio, r_p. The mean effective pressure is an indication of the work output and increases with pressure ratio for a fixed value of compression ratio. For an Otto cycle, the increase in the compression ratio results in an increase in the mean effective pressure and thermal efficiency.

5.5.2 The Diesel Cycle

In this cycle, the air is first compressed and then the fuel is injected into the cylinder and the combustion is initiated. Higher compression ratios can be achieved, and the air temperature should be higher than the self-ignition temperature of the fuel. In this engine no ignition system is required. The heat is added at constant pressure. Both in Otto cycle and the Diesel cycle, the heat is rejected at constant volume. Figures 5.5 (a) and (b) give the p-V and T-s diagram for a Diesel cycle.

The processes:

- 0–1 Intake
- 1–2 Compression-isentropic
- 2–3 Heat addition at constant pressure, Q_S
- 3–4 Expansion-isentropic-power stroke
- 4–1 Heat rejection at constant volume, Q_R
- 1–0 Exhaust

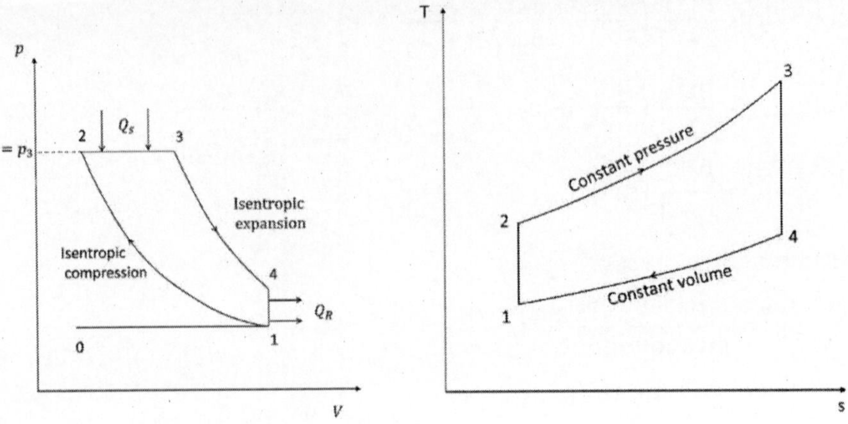

(a) (b)

Figure 5.5: The Diesel Cycle (a) p-V Diagram, (b) T-s Diagram

$$\frac{V_1}{V_2} = \text{Compression ratio} = r$$

$$\frac{V_3}{V_2} = \text{Cut} - \text{off ratio} = r_c$$

Thermal efficiency of the cycle,

$$\eta_{\text{Diesel}} = \frac{Q_S - Q_R}{Q_S}$$

$$= \frac{mC_p(T_3 - T_2) - mC_v(T_4 - T_1)}{mC_p(T_3 - T_2)}$$

$$= 1 - \frac{C_v}{C_p} \cdot \frac{(T_4 - T_1)}{(T_3 - T_2)}$$

$$\eta_{\text{Diesel}} = 1 - \frac{1}{\gamma}\left(\frac{T_4 - T_1}{T_3 - T_2}\right)$$

From isentropic relations, processes 1–2,

$$\frac{T_2}{T_1} = \left(\frac{V_1}{V_2}\right)^{(\gamma-1)}$$

$$= r^{(\gamma-1)}$$

For constant pressure process 2–3,

$$\frac{p_2 V_2}{T_2} = \frac{p_3 V_3}{T_3}$$

as, $p_2 = p_3$,

$$\frac{V_2}{T_2} = \frac{V_3}{T_3}$$

$$\frac{T_3}{T_2} = \frac{V_3}{V_2} = r_c \text{ (say)}$$

$$T_3 = T_2.r_c$$

therefore,

$$T_3 = T_1.r^{(\gamma-1)}.r_c$$

Process 3–4, Isentropic,

$$\frac{T_4}{T_3} = \left(\frac{V_3}{V_4}\right)^{(\gamma-1)}$$

$$T_4 = T_3\left(\frac{V_3}{V_2}.\frac{V_2}{V_4}\right)^{(\gamma-1)}$$

$$= T_3\left(\frac{r_c}{r}\right)^{(\gamma-1)} \quad (\,as\ V_4 = V_1)$$

therefore,

$$T_4 = T_1.r^{(\gamma-1)}.r_c.\left(\frac{r_c}{r}\right)^{(\gamma-1)}$$

$$= T_1.r_c^{\gamma}$$

therefore,

$$\eta_{\text{Diesel}} = 1 - \frac{1}{\gamma}\frac{T_1\left(\frac{T_4}{T_1} - 1\right)}{T_1\left(\frac{T_3}{T_1} - \frac{T_2}{T_1}\right)}$$

$$= 1 - \frac{1}{\gamma}\frac{(r_c^{\gamma} - 1)}{(r^{(\gamma-1)}.r_c - r^{(\gamma-1)})}$$

$$\eta_{\text{Diesel}} = 1 - \frac{1}{r^{(\gamma-1)}}\frac{(r_c^{\gamma} - 1)}{\gamma(r_c - 1))}$$

For a given compression ratio, the Otto cycle is more efficient. The Diesel cycle has a higher compression ratio, 16 to 20 than Otto cycle, where compression ratio is much lower 6–10. As for Diesel cycle the compression ratios are higher, the efficiency of a Diesel engine is more than that of the gasoline engine.

Work Output

The work output of a Diesel cycle,

$$W_N = p_2(V_3 - V_2) + \frac{p_3V_3 - p_4V_4}{\gamma - 1} - \frac{p_2V_2 - p_1V_1}{\gamma - 1}$$

and mean effective pressure, p_m

$$p_m = \frac{\text{work output}}{\text{swept volume}}$$

$$\text{Swept volume} = V_1 - V_2$$

5.5.3 Comparison of Otto and Diesel Cycles

Let us analyze the two cycles for the same compression ratio and heat addition. The p–V and T–s diagrams are given in Figure 5.6 (a) and (b).

$1 - 2 - 3 - 4 - 1$ Otto cycle

$1 - 2 - 3' - 4' - 1$ Diesel cycle

p_2/p_1 is same for both the cycles

As heat addition is same, From T–s diagram

Area $5 - 2 - 3 - 6$ = Area $5 - 2 - 3' - 6'$

for Otto cycle for Diesel cycle

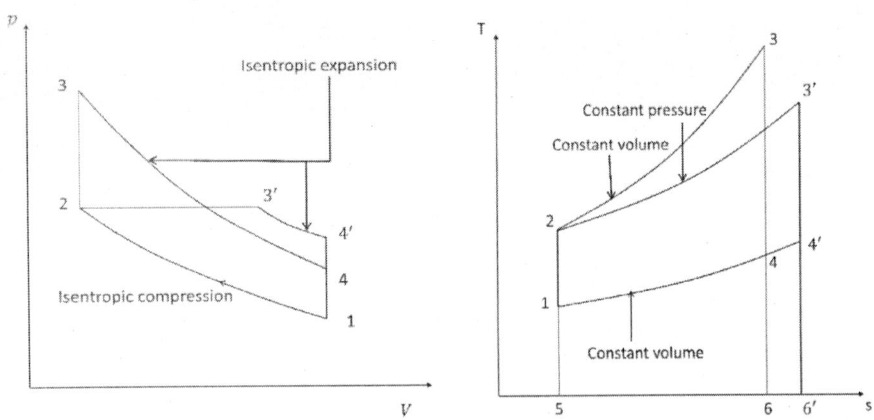

(a) (b)

Figure 5.6: (a) p–V and (b) T–s Diagrams for same Compression Ratio and Heat Addition

The heat rejection area 5-1-4-6 for Otto cycle is less than area 5-1-4′-6′ for Diesel cycle and therefore, the work output for Otto cycle will be more than the Diesel cycle. As the heat input is the same for both the cycles, the Otto cycle is more efficient than the Diesel cycle for the same compression ratio and energy input. Similarly, for other same input parameters also, the efficiency of both the cycles can be compared. It has been observed that the Otto cycle is always more efficient than the Diesel cycle.

5.6 CYLINDER ARRANGEMENTS

The reciprocating engines are classified by the cylinder arrangement. The term cylinder row means the arrangement where the centreline of the crankshaft journal is perpendicular to the plane of the centreline of the engine cylinders and the cylinder bank is the arrangement in which the centreline of the crankshaft journals is parallel to the plane containing the centreline of the engine cylinders.

A few cylinder arrangements are shown in Figure 5.7. The most common is inline cylinder arrangement where all cylinders are arranged linearly and transmit power to a single crankshaft. This is used in automobile engines. Four- and six-cylinder arrangements are popular. High powered automobiles use 8-cylinder V-engine. In this engine, there are two banks of cylinders i.e., two in-line engines inclined at an angle with one crankshaft. The opposed cylinder is arranged where two-cylinder banks are in the same plane on opposite sides of the crankshaft. It is a well-balanced engine and used in small aircraft.

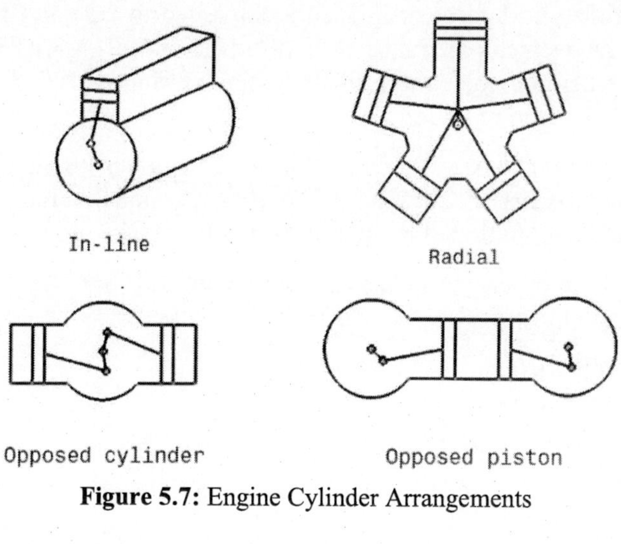

Figure 5.7: Engine Cylinder Arrangements

When a cylinder houses two pistons, each of which drives a separate shaft, is called an opposed piston engine. The opposed piston arrangement is used in large diesel installations.

The radial engines are used in conventional aircraft engine, where 3, 5, 7 or 9 cylinders are used in one bank and two to four banks are used. The odd number of cylinders is used from the point of view of balancing. The piston of all the cylinders is coupled to the same crankshaft.

5.7 CARBURETION

In spark ignition engine, a homogeneous mixture of fuel and air is prepared outside the engine cylinder. Carburetion is the process for mixing proper amount of fuel and air and the device is called carburetor. The various factors which influence the carburetion process are:

1. Engine speed,
2. fuel vaporization characteristics,
3. temperature of the incoming air and
4. the design of the carburetor.

Since the time available for carburetion is very short, e.g., for an engine running at 6000 rpm, it is 5 ms (milliseconds), the temperature and pressure of air has a large influence. Higher air temperature results in a decrease in power output of the engine.

The fuel-air mixture prepared may be stoichiometric, fuel rich or lean. We have seen in Chapter 4 that stoichiometric amount of air (i.e., air for complete combustion) for Octane, C_8H_{18} fuel is 15 kg/ kg fuel. Complete combustion means all carbon and hydrogen in fuel is converted to CO_2 and H_2O. When the air is less than stoichiometric i.e. A/F ratio 12:1 or 10:1, it is called fuel rich mixture and when the air is more than stoichiometric say, A/F ratio 17:1 or 20:1, it is called a lean mixture.

There is a limited range of A/F ratio within which combustion will occur. This limit ranges approximately between 9:1 to 19:1. Outside this range i.e., either too rich or too lean, stable flame propagation will not take place.

A modern carburetor has the following components:

1. Main metering system,
2. An idle system,
3. An enrichment system,
4. An accelerator pump,
5. A choke, and
6. Altitude compensation

An idling engine is one which operates at no load and with nearly closed throttle. Under these conditions, the engine requires a rich mixture. Figure 5.8 gives the representative pressure indicated during idling.

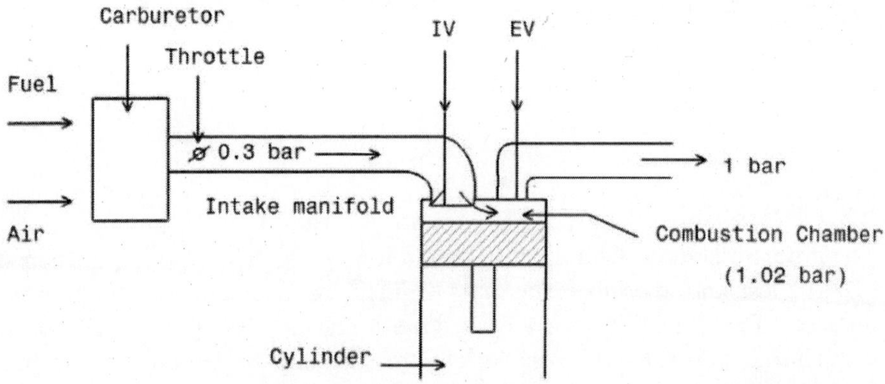

Figure 5.8: Combustion Chamber and Induction System at the Start of Intake Stroke

The amount of fresh charge in during idling is much less due to very small opening of the throttle. Overall, the combustion is poor and there is loss of power. As the throttle is gradually opened, the exhaust gas dilution of the fresh charge diminishes, and the mixture requirements become lean which is required for cruising operation.

A simple carburetor consists of a float chamber, fuel discharge nozzle and a metering orifice, a venturi, a throttle valve, and a choke. The float and needle valve maintain a constant level of gasoline in the float chamber. When the amount of fuel in the float chamber falls below the desired level, the float valve goes down opening the fuel supply valve and when the desired level has reached, the float closes the valve. In venturi, the velocity of air increases and pressure decreases.

5.7.1 Altitude Compensation

In conventional float type carburetor, the air and fuel are measured by volume and not on mass basis for calculation of air-fuel ratio. As the altitude increases the weight of air per cubic meter decreases due to change in density of air. Most of the carburetors are calibrated at sea level and tested under controlled environment. The mass of air at high altitude decreases and the mixture is fuel rich and, therefore, incomplete combustion takes place. The enrichment, E, due to variation in air density follows the relationship.

$$E + 1 = \sqrt{\rho_0/\rho}$$

where, subscript 'o' stands for calibrated conditions. Then if the density ratio is 0.84,

$$E + 1 = \sqrt{\frac{1}{0.84}}$$

$$= 1.091$$

$$E = 0.091$$

i.e., 9.1%

The problem is more with aircraft carburetors. At higher altitudes, the density of air is less and, therefore, the mass of intake air decreases and the power is reduced. The fuel flow must be reduced. The adjustment of fuel flow to compensate the changes in intake air pressure and temperature is the main function of mixture control.

Mixer control systems are classified as:

1. Back suction type, reducing the suction on the metering system.
2. Needle type, restricting the flow of fuel through the metering system, and
3. The air-port type, allowing the additional air to enter the carburetor.

5.8 FUEL INJECTION SYSTEMS

In fuel injection, the fuel speed at the point of delivery is greater that the airspeed to atomize the fuel. The injection is controlled by a pump which forces the fuel under pressure. The fuel is atomized into very fine particles. These droplets vaporise due to heat transfer from hot air to the fuel, the temperature reaches a value higher than its self-ignition temperature. The fuel is injected into the combustion chamber towards the end of compression stroke and the fuel ignites spontaneously initiating the combustion process.

The requirements of an injection system are:

1. accurate metering of fuel per cycle,
2. timing the injection of fuel,
3. proper control of rate of injection,
4. proper atomization of fuel into fine droplets,
5. proper spray pattern for rapid mixing,
6. uniform distribution of fuel droplets,
7. equal quantity of fuel to all cylinders in a multi cylinder engine, and
8. no time lag during beginning and end of injection.

Atomization is done by forcing the fuel through a small orifice under high pressure. The injector assembly consists of:

1. a needle valve,
2. a compression spring,
3. a nozzle, and
4. an injector body.

Through nozzle, the liquid fuel is sprayed into the combustion chamber. The functions of a nozzle are:

1. atomization,
2. distribution of fuel which depends upon injection pressure, density of air in the cylinder and the physical properties of fuel,
3. prevention of impingement on walls, and
4. mixing.

In a gasoline injection system, the fuel is injected into the intake manifold or near the intake port through an injector. The position of the throttle valve controls the flow of the mixture into the intake manifold. In a carburetor engine, uniformity of mixture is difficult in each cylinder of a multi cylinder engine.

By adopting gasoline injection, each cylinder can get the same richness of the air gasoline mixture and the malnutrition can be avoided to a great extent.

The multipoint fuel injection (MPFI) provides proper ratio of gasoline and air to the cylinders. There are two types of arrangements. In the port injection, the injector sprays gasoline into the air, inside the intake manifold. This mixture passes through the intake valve and enters the cylinder. Every cylinder is provided with an injector in its intake manifold. Thus for 6 cylinders there will be 6 injectors.

In a single point throttle body injection system, the throttle body is similar to the carburetor body, with the throttle valve controlling the amount of air entering the intake manifold. An injector is placed above the throat of the throttle body. The injector sprays gasoline into the air in the intake manifold where the gasoline mixes with air. The mixture then passes through the throttle valve and enters the intake manifold.

5.9 IGNITION

An ignition system is required in a spark ignition engine to ignite the fuel air mixture for initiation of combustion. The electrical discharge produced between the two electrodes of a spark plug starts the combustion process. The combustion is initiated close to the end of compression stroke. The spark is produced in a repeatable manner i.e., cycle by cycle over the full range of load and speed of the engine. A spark can be caused by applying high voltage between two electrodes separated by a gap. There is a critical voltage below which no sparking occurs.

With a homogeneous mixture in cylinder, spark energy of the order of 1 mJ and a duration of a few microseconds may be enough to initiate combustion. The pressure, temperature and density of the mixture have an influence on the voltage required to produce the spark. Usually, spark energy of 40 mJ and the duration longer than 0.5 ms is required for reliable ignition.

5.10 COMBUSTION CHAMBER

The chemical reactions between carbon and hydrogen in fuel with oxygen in air is called combustion and results in liberation of heat and increase in temperature of the product gas. In spark ignition engine a homogeneous mixture is formed in the carburetor. The combustion is initiated towards the end of compression stroke. The flame front moves through the mixture at a certain velocity. For a stoichiometric mixture of gasoline-air, the flame speed is about 40 cm/s. In S.I. engine, the maximum flame speed is obtained at equivalence ratio, ϕ between 1.1 to 1.2, i.e., for a slightly fuel rich mixture.

Typical theoretical pressure-crank angle (p-θ) diagram for a S.I. engine is given in Figure 5.9. For an ideal four stroke engine, a–b is compression, b–c is combustion and c–d is expansion process. In theoretical process, the entire pressure rise during combustion is at constant volume, process b–c i.e., at top dead center (TDC). In actual process, practically the steps are shown in Figure 5.10. The spark plug is located before the end of compression at A and flame propagation begins. Before TDC, at B the motoring starts. The pressure rises due to combustion and the peak pressure is reached at C.

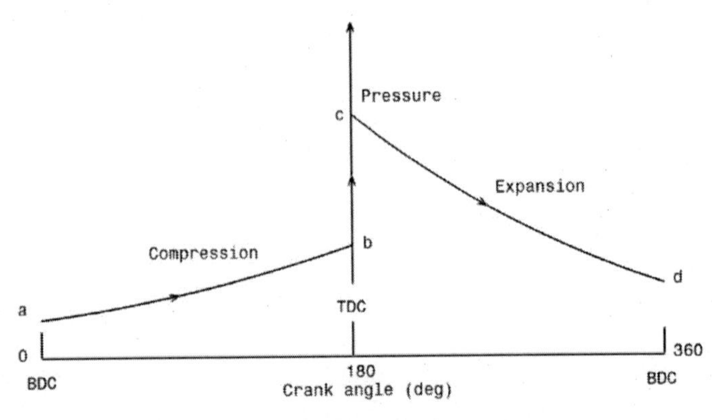

Figure 5.9: Theoretical (p–θ) Diagram

In normal combustion, the flame front moves in a uniform manner. The flame front advances through the mixture and the temperature and pressure of the unburned mixture increases. If the temperature of the unburned mixture exceeds the self-ignition temperature of the fuel and remains at or above this temperature during the period of pre-flame reactions, spontaneous ignition occurs at various locations. The phenomenon is called knocking.

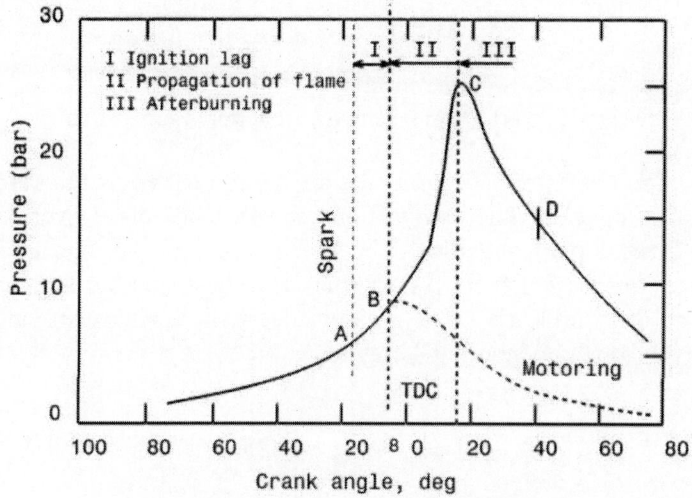

Figure 5.10: Combustion in a S.I. Engine

In a spark ignition engine, the combustion is initiated by the spark plug and the flame front moves across the fresh mixture to the other end of the combustion chamber. The heat released increases the temperature and pressure of the burning mixture. Figure 5.11 (a) shows this phenomenon. Figure 5.11 (b) shows the combustion process with knock. In this figure, a part of the mixture in zone $CC'D$ gets ignited early and sudden increase in temperature and pressure take place. The phenomenon of knocking results in loss of power and damage to the engine.

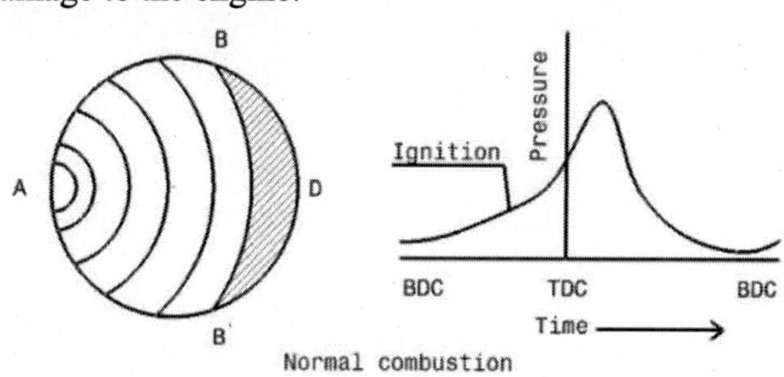

(a)

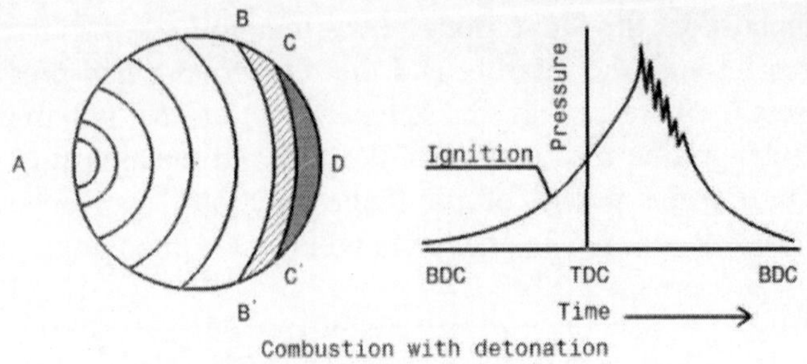

(b)

Figure 5.11: (a) Normal and (b) Abnormal Combustion

In compression ignition (CI) engine, the air is compressed and in compressed air the fuel is injected. The combustion process takes place over a definite interval of time. If there is ignition delay and more droplets of fuel are injected, and a greater quantity of fuel gets accumulated, the actual burning will result in rapid rate of pressure rise. The pressure rise will be almost instant. Figure 5.12 shows the phenomenon in CI engine.

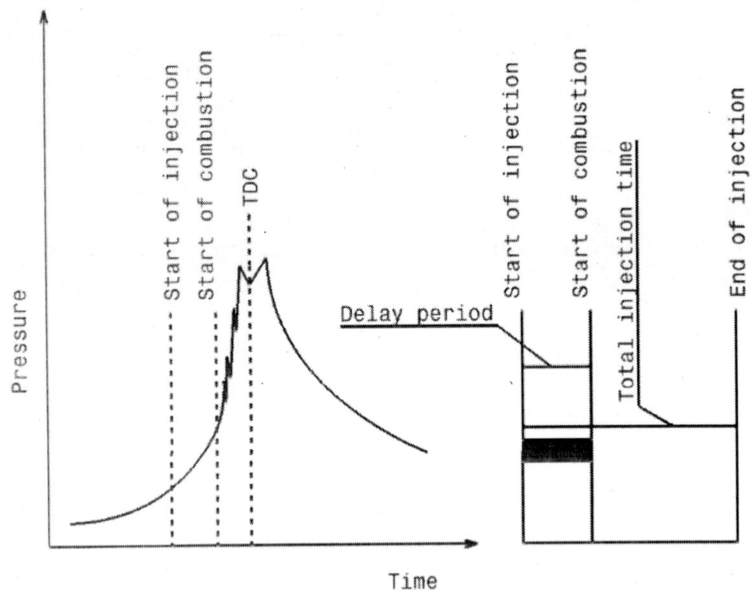

Figure 5.12: Abnormal Combustion in C.I. Engine

A notable difference in C.I. and S.I. engine knocking phenomenon is that, in S.I. engine it takes place in the later part of combustion process, while in C.I. engine, it takes place in the early part of the combustion process.

5.11 ENGINE COOLING AND LUBRICATION

In an I.C. engine, only 25–35% of chemical energy present in the fuel is converted to mechanical energy and about 35% is lost through the cooling systems and the remainder goes through exhaust and lubrication. The combustion chamber temperature goes to a level as high as 2000 K. Therefore, proper cooling of the walls of the combustor is an essential requirement. Even the lubrication oil undergoes both physical and other changes. High cylinder wall or head temperature may lead to auto ignition in S.I. engines. Therefore, the heat of the walls of the combustion chamber is continuously removed by a suitable cooling system. Heat lost through lubrication oil and radiation is about 5% of the total heat supplied.

There are two types of cooling systems:

1. Liquid cooling,
2. Air cooling

In liquid cooling mainly water is used and is circulated through the jackets provided around the cylinder, cylinder head, valve ports and seats where it extracts most of the heat. The heat is transferred from the cylinder walls through convection and conduction. The liquid is heated in passage and is cooled by an air-cooled radiator system.

In an air-cooled system, air is made to flow past the outside of the cylinder barrel and the fins are provided to increase the surface area. The rate of cooling is better in this system. The cylinder wall temperature of the air-cooled system is considerably higher than those of water-cooled systems.

Friction is defined as forces acting between surfaces in relative motion. The frictional losses are due to sliding as well as rotating parts. The difference between indicated power and brake power is called frictional power. The frictional power loss is about 30% in reciprocating engines. The frictional power loss and wear of the parts is controlled by the proper lubrication.

Lubrication is necessary to reduce friction and wear between the components in an engine. The lubricating system has the following functions:

1. To reduce the frictional resistance to give maximum mechanical efficiency.
2. To protect the engine parts against wear.
3. To cool the piston and other regions in the engine.
4. To remove all impurities from lubricated regions.
5. To hold the gas and oil leakage within acceptable limits.

The lubricating oil is injected into the engine through a pressurised system. The oil pump drains oil from the engine sump and delivers it through a control valve to the oil cooler. The oil then passes through the filter to the main oil gallery. From the main gallery it is branched to the main, the big end, and the crankshaft bearings. Oil is also diverted to the oil pump.

The various lubricating systems used in internal combustion engines are:

1. Mist lubrication system
2. Wet sump lubrication system
3. Dry sump lubrication

In two stroke engines, the mist lubrication system is used. The lubricating oil is mixed with the fuel usually 3–6% and the oil-fuel mixer is inducted through the carburetor. The fuel is vaporised and the oil in the form of mist goes via the crankcase into the cylinder. The oil which strikes the crankcase walls lubricate the main and the connecting rod bearings. The rest of the oil lubricates the piston, piston rings and the cylinder.

In a wet sump system, the bottom of the crankcase contains an oil pan or sump from which the lubricating oil is pumped to various engine components by a pump. After lubricating these parts, the oil flows back to the sump by gravity.

In a dry sump lubricating system, the supply oil is carried in an external tank. An oil pump draws oil from the supply tank and circulates it under pressure to the various bearings of the engine.

The lubricant should have the following properties:

1. Suitable viscosity
2. Oiliness to ensure adherence to the bearings.
3. High strength to prevent metal to metal contact
4. Should not react with the lubricating surfaces
5. Non-toxic, non-inflammable and non-foaming characteristics.
6. Low cost.

The modern lubricant for heavy duty is highly refined and for these reasons the lubricants are seasoned by the addition of certain oil soluble organic compounds containing inorganic elements such as phosphorus, sulphur, amine derivatives. Metals are added to the mineral based lubricating oil to exhibit the desired properties.

The SAE (Society of Automotive Engineers) has developed standards for grading lubricating oils. They are defined in SAE J300 standard.

5.12 ENGINE PERFORMANCE

The performance is indicated by the term efficiency, η. Various types of efficiency parameters are as follows:

1. Indicated thermal efficiency, $\eta_{i\,th}$ is the ratio of energy in the indicated power, i.p. to the input fuel energy.

$$\eta_{i\,th} = \frac{\text{i.p (kJ/s)}}{\text{heating value of the fuel (kJ/s)}}$$

Mean effective pressure (p_m) is the average pressure inside the cylinders of an engine based on power output. For a particular engine, there will be indicated mean effective pressure, *imep* (p_{im}) and corresponding brake mean effective pressure, *bmep* (p_{bm}). The p-V diagram for an ideal four stroke Otto cycle showing *imep* is given in Figure 5.13.

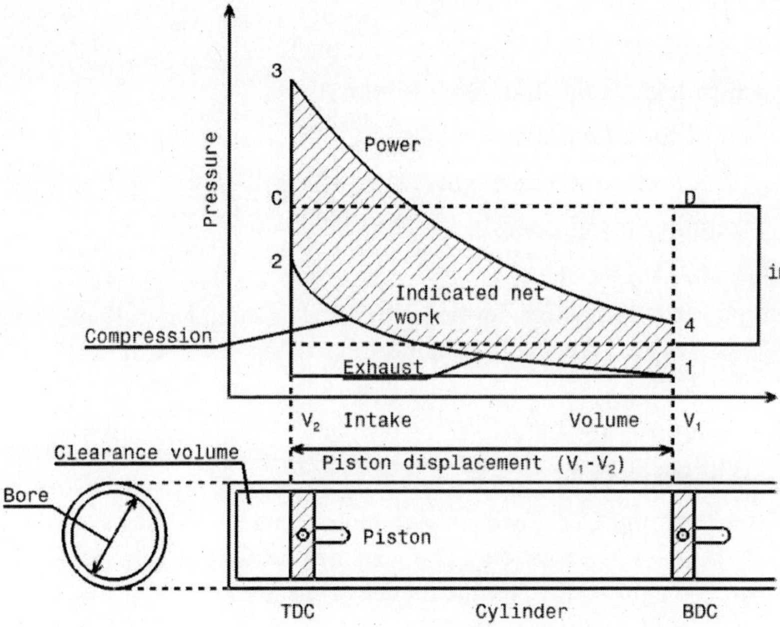

Figure 5.13: p-V Diagram for an Ideal Four Stroke Otto Cycle

(*Source:* Ganesan V., Internal Combustion Engines, McGraw Hill Education (India) Private Limited, New Delhi, Fourth Edition, 2013)

$$p_{\text{im}} = \frac{\text{Net work of cycle}}{V_1 - V_2}$$

$$= \frac{\text{Area of indicator diagram}}{\text{Length of indicator diagram}}$$

Power is defined as the rate of doing work. Indicated power, ip

$$ip = \dot{m}_a \times \text{net work}$$

where, $\dot{m}_a$ = mass flow rate of air, kg/s

Net work is in kJ/kg of air and ip is in kW.

$$\text{indicated net work per cycle} = p_{\text{im}} \times V_s$$

where, V_s is swept volume $(V_1 - V_2)$:

$$ip = \frac{p_{im}.\,L.\,A.\,n.\,K}{1000 \times 60}$$

where, ip = indicated power, kW

p_{im} = indicated mean effective pressure, N/m^2

L = length of stroke, m

A = area of piston, m^2

N = speed, revolutions per minute (rpm)

n = number of power stroke per minute

= $N/2$ for a four stroke engine

= N for a two stroke engine

K = number of cylinders.

2. *Brake Power (bp)*

Indicated power is the power developed within the cylinder and brake power (bp) is the power delivered by the engine.

$$bp = \frac{2\pi NT}{60000}\ kW$$

where, T is torque in N-m.

Torque is the uniform or fluctuating turning moment or twist exerted at a distance from the axis of rotation. For an engine operating at a given speed and delivering a given power, the torque remains constant.

$$ip = bp + fp$$

where, fp = friction power

The ratio of power developed to the maximum usable power at the same speed is called the load. The specific fuel consumption varies with load and speed. The performance is the relationship between power developed, speed and specific fuel consumption at operating conditions within useful range of speed and load.

3. *Mechanical efficiency,* η_m

The mechanical efficiency is the ratio of brake power to indicated power.

$$\eta_m = \frac{bp}{ip}$$

4. *Combustion efficiency,* η_b

It is the ratio of heat liberated to the theoretical heat available in the fuel. The heat liberated is always less than the heating value due to incomplete combustion. The combustion efficiency is more than 90%. In a typical I.C. engine, the energy available to the brake is about 20% and the rest is lost to the coolant, exhaust and to the surroundings.

5.13 SUPERCHARGING

The aim of an engine designer is to get improved power output with minimum exhaust emissions. The power output can be increased by increasing the mean effective pressure. Air or air-fuel mixture is supplied at a pressure higher than the atmospheric pressure. This will increase density, thereby mass of air or air-fuel mixture for the same swept volume. This is called supercharging. Increased density will help in combustion as more oxygen is available. Therefore, more fuel can be inducted, and higher power is achieved.

Supercharger is a pressure boosting device which supplies air or mixture at a higher pressure. A centrifugal or axial flow or displacement type compressor

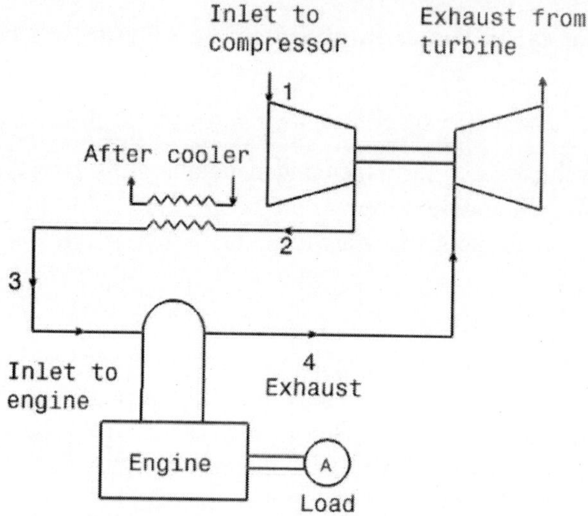

Figure 5.14: A turbo-supercharger Arrangement

is normally used. Some superchargers are driven by a gas turbine. The turbine can be driven by engine exhaust gases. Such a supercharger is called turbocharger. Figure 5.14 gives the schematic diagram of a typical gas turbine supercharging arrangement using engine exhaust gas. In this arrangement exhaust energy of the engine is used to drive the turbine which is coupled to a compressor. An aftercooler is also shown by which the cool air to the engine can be delivered. This will further increase the density of the intake air.

5.14 ENGINE DESIGN

Engine ratings indicate the highest power to give satisfactory economy, reliability, and durability under service conditions. Maximum torque and the speed at which it is achieved is also given. The various parameters are:

1. Mean piston speed—indicates the success in handling loads, resistance to airflow and engine friction.
2. Brake mean effective pressure—indicates the ability to induct air, fuel air ratio and fuel conversion efficiency. In supercharged engine, it indicates the success of handling higher gas pressures and thermal loading.
3. Power per unit area—indicates effectiveness of use of piston area
4. Specific weight—indicates economy in use of materials
5. Specific volume—indicates effectiveness of engine space
6. Engine speed—for use with full throttle or with maximum fuel pump setting
7. Brake specific fuel consumption or fuel conversion efficiency
8. Brake specific emissions.

The four-stroke cycle dominates except in the smallest and largest engine sizes.

The maximum mean piston speed is in the range of 8–15 m/s.

For airborne vehicles i.e., airplanes and helicopters, four stroke S.I. engine with air cooling and engine power in the range of 45–2700 kW for airplanes and 45 to 1500 kW for helicopters are used.

SUMMARY

An internal combustion (I.C.) engine is a reciprocating engine, where the piston moves back and forth in a cylinder and transmits power through a connecting rod and crank mechanism to drive the shaft. In these engines, the chemical energy of fuel is converted to thermal energy, which then produces mechanical power. The fuel is burned inside the engine. The hot products of combustion are the working fluid to produce mechanical energy. Nicolaus A.

Otto proposed an engine with four strokes: intake, compression before ignition, expansion or power stroke and exhaust stroke. Four-stroke engine produces one power stroke in two revolutions while a two-stroke engine produces one power stroke in every revolution.

Rudolf Diesel developed the compression engine, where the air was compressed, and the liquid fuel was injected in hot compressed air and combustion was initiated. This has resulted in greater expansion without detonation or knock increase and improved engine efficiency. In S.I. Engines, the pre-mixed fuel and air using a carburetor or a fuel injection system is introduced into the cylinder. In C.I. engines, only air is introduced into the cylinder and compressed.

The thermal efficiency of an ideal four-stroke Otto cycle is given as,

$$\eta_{otto} = 1 - \frac{1}{r^{(\gamma-1)}}$$

where, r is compression ratio.

This shows that efficiency of the cycle is a function of compression ratio, r and specific heat ratio, γ only. For a given compression ratio, the Otto cycle is more efficient.

The Diesel cycle has a higher compression ratio. As the compression ratios are higher, the efficiency of a Diesel cycle is more than that of a gasoline engine.

The reciprocating engines are classified by the cylinder arrangements. The radial engines are used on small trainer aircraft engine. The odd number of cylinders are used from the point of view of balancing. The piston of all the cylinders is connected to the same crankshaft.

Carburetion, fuel injection systems, ignition and combustion chamber are discussed. The abnormal combustion due to knocking in both S.I. and C.I. engines is not desirable. In normal combustion, the flame front moves in a uniform manner. If the temperature of the unburned mixture exceeds the self-ignition temperature of the fuel and remains at or above this temperature during the period of pre-flame reactions, spontaneous ignition occurs at various locations. This phenomenon is called knocking. Engine cooling and lubrication aspects are also discussed.

The performance of an engine is assessed by indicated thermal efficiency, brake power, mechanical efficiency, etc. To increase the power output, supercharging and turbocharging is used and are discussed. Some design aspects are also given.

EXAMPLES

Example 5.1

Estimate the efficiency of an Otto cycle having a compression ratio of 6, when the C_v increases by 1.4%.

Solution

$$\eta_{otto} = 1 - \frac{1}{r^{(\gamma-1)}}$$

$$C_p - C_v = R;$$

$$C_p/C_v = \gamma;$$

$$\gamma - 1 = R/C_v$$

$$1 - \eta = r^{-R/C_v}$$

taking ln,

$$ln(1-\eta) = -(R/C_v)\, ln\, r$$

differentiate,

$$-\frac{1}{(1-\eta)} d\eta = \left(\frac{R}{C_v{}^2}\right) \ln r\, (dC_v)$$

$$d\eta = -\frac{(1-\eta)}{C_v{}^2} R \ln r\; dC_v$$

$$\frac{d\eta}{\eta} = -\frac{(1-\eta)}{\eta\, C_v{}^2} (C_p - C_v) \ln r\; dC_v$$

$$\frac{d\eta}{\eta} = -\frac{(1-\eta)}{\eta} (\gamma - 1) \ln r\, \frac{dC_v}{C_v}$$

$$\eta = 1 - \left(\frac{1}{6}\right)^{0.4}$$

$$= 0.5116$$

$$= 51.16\%$$

$$\frac{d\eta}{\eta} = -\frac{1 - 0.5116}{0.5116} (1.4 - 1)(\ln 6). \frac{1.4}{100}$$

$$= -0.0096$$

$$= \mathbf{-0.96\,\%}$$

Example 5.2

Estimate the brake specific fuel consumption of a Diesel engine. The fuel consumption is 4 grams per second when the power output is 60 kW. If the mechanical efficiency is 80%, calculate the indicated specific fuel consumption.

Solution

$$bsfc = \frac{\dot{m}_f}{bp} = \frac{4}{60}$$

$$= 0.066\ g/kW.s$$

$$= (0.066/1000) \times 3600$$

$$bsfc = \mathbf{0.2376\ kg/kW.h}$$

$$isfc = bsfc.\eta_m$$

$$= 0.2376\ \times 0.8$$

$$= \mathbf{0.19008\ kg/kW.h}$$

QUESTIONS

1. State the assumptions made for analyzing an air standard cycle. Differentiate between S.I. engine and C.I. engine.
2. Discuss the difference between four-stroke and two-stroke engines. Explain the working of a four-stroke S.I. engine with p-V and T-s diagram.
3. Analyze and discuss the p-V and T-s diagram of a diesel engine. Also, derive the expression for thermal efficiency.
4. With the help of neat diagrams, explain the various types of cylinder arrangements possible in an engine.
5. Write short notes on:
   (a) Carburetion
   (b) Knocking phenomena
   (c) Supercharging

CHAPTER 6

Gas Turbines

6.1 INTRODUCTION

A simple gas turbine consists of a compressor, combustion chamber and a turbine (Figure 1.4). Both, heat addition and heat rejection processes are at constant pressure. The constant pressure systems have the advantage to handle higher air mass flows, which leads to generation of high power. In aircraft propulsion systems, the performance depends significantly upon forward speed and altitude. The variations in intake air conditions i.e. mass flow rate, density and pressure are very wide. It is better power/weight ratio, which has enabled gas turbines to find use over reciprocating engines. Except for low powered light aircraft, almost all modern engines use gas turbine to generate thrust power. The useful power output is in the form of thrust, which is generated in a propelling nozzle.

6.2 IDEAL CYCLE

The ideal cycle for simple gas turbine is Brayton or Joule cycle. The $p - V$ and $T - s$ diagrams for the cycle are given in Figure 6.1 (a) and (b) respectively.

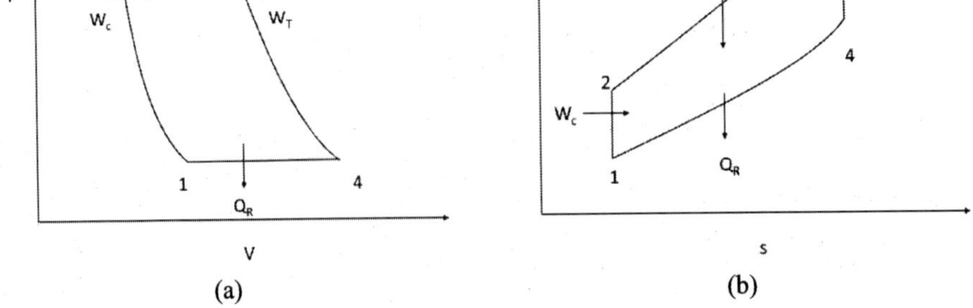

(a) (b)

Figure 6.1: (a) $p - V$ Diagram, (b) $T - s$ Diagram

(*Source:* Ganesan V., Gas Turbines, Tata McGraw Hill Publishing Company Limited, Second Edition, 2007)

To analyse the cycle, the following assumptions are made:

1. The compression and expansion processes are isentropic, i.e. reversible and adiabatic.
2. The change in kinetic energy of the working fluid between inlet and outlet of each component is negligible.
3. There are no pressure losses in combustion chamber and ducts connecting each component.
4. The working fluid composition does not change throughout the cycle and is a perfect gas with constant specific heats.
5. The mass flow of fluid remains constant throughout the cycle.
6. The heat transfer losses during the cycle are neglected.
7. Friction losses are neglected.

The various steps in the cycle are:

1–2 Compression, air pressure and temperature increases .

2–3 Heat addition in combustion chamber at constant pressure by burning fuel. Temperature increases.

3–4 Expansion of hot products of combustion in turbine, generation of power. Pressure and temperature of gases decrease.

4–1 Heat rejection at constant pressure.

Compressor work,

$$W_c = h_2 - h_1 = C_p(T_2 - T_1)$$

Turbine work,

$$W_T = h_3 - h_4 = C_p(T_3 - T_4)$$

Heat addition,

$$\begin{aligned} Q_S &= h_3 - h_2 \\ &= C_p(T_3 - T_2) \end{aligned}$$

Heat rejection,

$$\begin{aligned} Q_R &= h_4 - h_1 \\ &= C_p(T_4 - T_1) \end{aligned}$$

Net work output,

$$\begin{aligned} W_N &= W_T - W_C \\ &= C_p(T_3 - T_4) - C_p(T_2 - T_1) \\ &= C_p\left(\frac{T_3}{T_1} - \frac{T_4}{T_1} - \frac{T_2}{T_1} + 1\right)T_1 \end{aligned}$$

Let,

$$\frac{T_3}{T_1} = t, \frac{p_2}{p_1} = r$$

where, t is maximum to minimum temperature ratio in the cycle and r is compressor pressure ratio

From isentropic relations,

$$\frac{T_2}{T_1} = \left(\frac{p_2}{p_1}\right)^{\frac{(\gamma-1)}{\gamma}} = r^{(\gamma-1)/\gamma}$$

$$\frac{T_3}{T_4} = \left(\frac{p_3}{p_4}\right)^{\frac{(\gamma-1)}{\gamma}} = \left(\frac{p_2}{p_1}\right)^{\frac{(\gamma-1)}{\gamma}} = r^{(\gamma-1)/\gamma}$$

(as, $p_3 = p_2, p_4 = p_1$)

therefore,

$$\frac{T_2}{T_1} = \frac{T_3}{T_4}$$

$$\frac{T_4}{T_1} = \frac{T_3}{T_2}$$

$$W_N = C_p T_1 \left(t - \frac{T_4}{T_3} . \frac{T_3}{T_1} - \frac{T_2}{T_1} + 1\right)$$

$$= C_p T_1 \left(t - \frac{1}{r^{(\gamma-1)/\gamma}} . t - r^{(\gamma-1)/\gamma} + 1\right)$$

cycle efficiency,

$$\eta = \frac{\text{Net work output}}{\text{Heat input}} = \frac{Q_S - Q_R}{Q_S}$$

$$= \frac{C_p(T_3 - T_4) - C_p(T_2 - T_1)}{C_p(T_3 - T_2)}$$

$$= \frac{(T_3 - T_2) - (T_4 - T_1)}{(T_3 - T_2)}$$

$$= 1 - \frac{T_4 - T_1}{T_3 - T_2}$$

$$= 1 - \frac{T_1\left(\frac{T_4}{T_1} - 1\right)}{T_2\left(\frac{T_3}{T_2} - 1\right)}$$

$$= 1 - \frac{T_1}{T_2}$$

$$\left(\text{as}, \frac{T_4}{T_1} = \frac{T_3}{T_2}\right)$$

$$\eta = 1 - \frac{1}{(T_2/T_1)}$$

$$\eta = 1 - \frac{1}{r^{(\gamma-1)/\gamma}}$$

This shows that the efficiency of the cycle depends on the pressure ratio and the nature of gas, γ only. A typical plot of net specific work out, $\frac{W_N}{C_p T_1}$ and efficiency of the cycle with respect to pressure ratio, r are given figure 6.2 (a) and 6.2 (b) respectively.

It may be observed that the work output increases with increase in $t = T_3/T_1$ and efficiency of the cycle increases with higher specific heat ratio, γ of the working fluid. The term t depends on maximum cycle temperature T_3, the metallurgical limit. For aircraft application, a value of $t =$ 5 to 5.5 is permissible with cooled turbine blades. For maximum power output, differentiate $\frac{W_N}{C_p T_1}$ w.r.t $r^{(\gamma-1)/\gamma}$ and put equal to zero.

$$r_{opt}^{(\gamma-1)/\gamma} = \sqrt{t}$$

$$t = \left(r_{opt}^{(\gamma-1)/\gamma}\right)^2 = \frac{T_2}{T_1}.\frac{T_3}{T_4} = \frac{T_2}{T_4}.\frac{T_3}{T_1} = \frac{T_2}{T_4}.t$$

$$T_2 = T_4$$

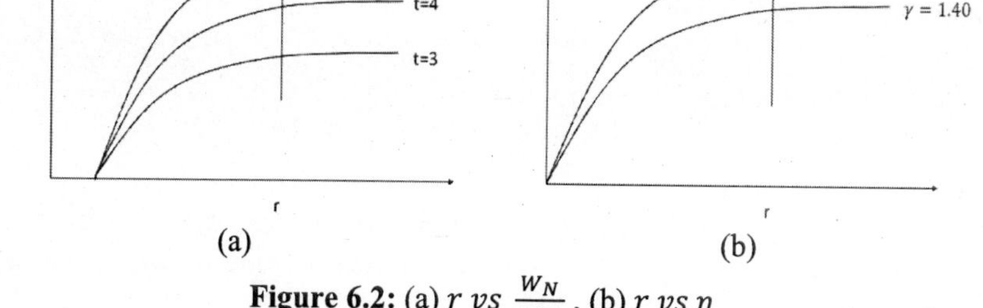

Figure 6.2: (a) r vs $\frac{W_N}{C_p T_1}$, (b) r vs η

(*Source:* Ganesan V., Gas Turbines, Tata McGraw Hill Publishing Company Limited, Second Edition, 2007)

This shows that the specific work output of the system is maximum when the pressure ratio is such that the compressor and turbine outlet temperatures are equal. Figure 6.2 (b) shows that the efficiency of the system will improve when a monoatomic gas such as argon ($\gamma = 1.66$) than air ($\gamma = 1.40$) is used as working fluid.

6.3 STAGNATION PROPERTIES

Stagnation properties take care of the kinetic energy term in steady flow energy equation. Stagnation enthalpy, h_0 is the enthalpy which a gas stream of enthalpy, h and velocity, V would possess when brought to rest adiabatically, $(V=0)$, and without work transfer. The energy equation then becomes,

$$h_0 + 0 = h + \frac{V^2}{2}$$

$$h_0 = h + \frac{V^2}{2}$$

When the fluid is a perfect gas,

$$h = C_pT, h_0 = C_pT_0$$

$$C_pT_0 = C_pT + \frac{V^2}{2}$$

$$T_0 = T + \frac{V^2}{2C_p}$$

The term, $\dfrac{V^2}{2C_p}$ is called dynamic temperature. T is referred to as static temperature and T_0 stagnation temperature.

An idea of the order of magnitude of the difference between T_0 and T is obtained by considering air at atmospheric temperature, $C_p = 1.005$ kJ/ kg. K flowing at 100 m/s. Then

$$T_0 - T = \frac{100^2}{2 \times 1.005 \times 10^3}$$

$$\approx 5\ K$$

It follows from the energy equation that if there is no heat or work, T_0 will remain constant. Applying the concept of adiabatic compression , the energy equation becomes,

$$W = -C_p\left[(T_2 - T_1) + \frac{V_2^2}{2C_p} - \frac{V_1^2}{2C_p}\right] = -C_p(T_{02} - T_{01})$$

Similarly, for heating process, without work transfer,

$$Q = C_p(T_{03} - T_{02})$$

Thus, if stagnation properties are employed there is no need to refer to the kinetic energy term. It may be noted that, it is easier to measure the stagnation temperature of a high velocity stream than the static temperature. In the real cycle analysis, stagnation conditions are referred and not static conditions as in the ideal cycle analysis.

6.4 COMPRESSOR AND TURBINE EFFICIENCIES

The compressor efficiency may be defined as the ratio of the work required for isentropic compression to the actual work input. Figure 6.3 gives the compression process in real system taking into consideration stagnation/total properties.

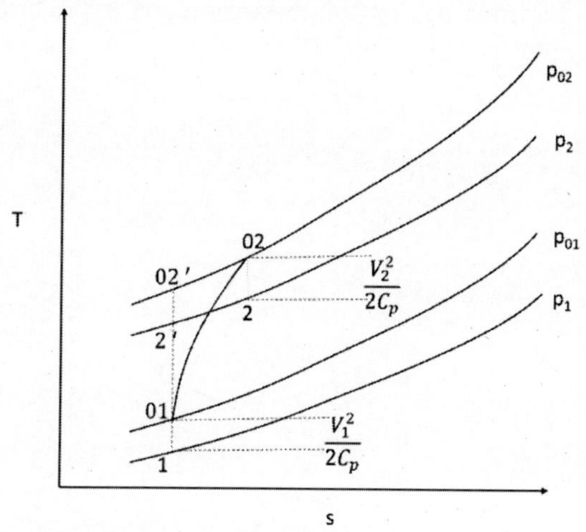

Figure 6.3: Stagnation and Static Properties in Compression Process

compression efficiency,

$$\eta_c = \frac{\text{isentropic compression work}}{\text{actual compression work}}$$

$$= \frac{h_{02'} - h_{01}}{h_{02} - h_{01}}$$

$$= \frac{C_p(T_{02'} - T_{01})}{C_p(T_{02} - T_{01})}$$

$$= \frac{(T_{02'} - T_{01})}{(T_{02} - T_{01})}$$

The ratio of stagnation pressure at outlet to the stagnation pressure at inlet during compression process is r_c. Then from isentropic relations, we have

$$\frac{T_{02'}}{T_{01}} = \left(\frac{p_{02}}{p_{01}}\right)^{\frac{(\gamma-1)}{\gamma}}$$

$$= r_c^{(\gamma-1)/\gamma}$$

therefore,

$$\eta_c = \frac{T_{01}\left(\frac{T_{02'}}{T_{01}}-1\right)}{(T_{02}-T_{01})}$$

$$(T_{02} - T_{01}) = \frac{T_{01}}{\eta_c}\left[r_c^{(\gamma-1)/\gamma} - 1\right]$$

From this equation, the actual temperature rise during compression process can be calculated. Figure 6.4 gives the expansion process in turbine considering stagnation properties.

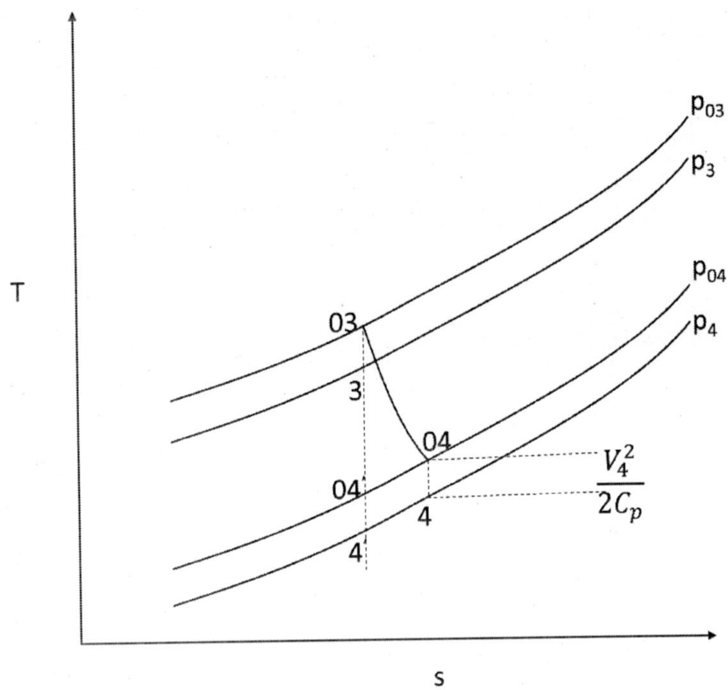

Figure 6.4: Stagnation and Static Properties in Expansion Process

The efficiency of a gas turbine,

$$\eta_T = \frac{\text{actual turbine work output}}{\text{isentropic turbine work output}}$$

$$= \frac{h_{03} - h_{04}}{h_{03} - h_{04'}}$$

$$= \frac{C_p(T_{03} - T_{04})}{C_p(T_{03} - T_{04'})}$$

$$= \frac{(T_{03} - T_{04})}{(T_{03} - T_{04'})}$$

Let $\frac{p_{03}}{p_{04}} = r_t$

From isentropic relations,

$$\frac{T_{03}}{T_{04'}} = \left(\frac{p_{03}}{p_{04}}\right)^{\frac{(\gamma-1)}{\gamma}}$$

$$= r_t^{(\gamma-1)/\gamma}$$

therefore,

$$(T_{03} - T_{04}) = \eta_T(T_{03} - T_{04'})$$

$$= \eta_T . T_{03}\left(1 - \frac{T_{04'}}{T_{03}}\right)$$

$$= \eta_T . T_{03}\left(1 - \frac{1}{T_{03}/T_{04'}}\right)$$

$$= \eta_T . T_{03}\left(1 - \frac{1}{r_t^{(\gamma-1)/\gamma}}\right)$$

The net work done,

$$W_N = W_T - W_C$$

$$= C_p(T_{03} - T_{04}) - C_p(T_{02} - T_{01})$$

$$= C_p . \eta_T . T_{03}\left(1 - \frac{1}{r_t^{(\gamma-1)/\gamma}}\right) - C_p . \frac{T_{01}}{\eta_{c.}}\left(r_c^{(\gamma-1)/\gamma} - 1\right)$$

If works of compression and expansion are equated, the minimum temperature ratio $\frac{T_{03}}{T_{01}} = t_{min}$ can be calculated for which no net work output is possible.

Thus,

$$C_p.\eta_T.T_{03}\left(1-\frac{1}{r_t^{(\gamma-1)/\gamma}}\right)=C_p.\frac{T_{01}}{\eta_{c.}}\left(r_c^{(\gamma-1)/\gamma}-1\right)$$

Assuming, the specific heats in both compression and expansion are same and also,

$$r_c = r_t = r$$

$$\frac{T_{03}}{T_{01}} = \frac{r^{(\gamma-1)/\gamma}}{\eta_T.\eta_c}$$

therefore,

$$t_{min} = \frac{T_{03}}{T_{01}} = \frac{r^{(\gamma-1)/\gamma}}{\eta_T.\eta_c}$$

Thus for $r = 5$, $\eta_c = \eta_T = 0.85$, $\gamma = 1.4$; the minimum temperature ratio is about 2.19. Therefore, with $T_{01} = 300$ K, T_{03} must be about 657 K. Hence, T_{03} must be higher than this value to have any power output. Due to low efficiencies of compressor and turbine and non availability of materials suitable for high temperature involved in turbine, gas turbine could not find application till 1930's.

6.5 ACTUAL GAS TURBINE CYCLE

Figure 6.5 gives the actual or real gas turbine cycle with losses.

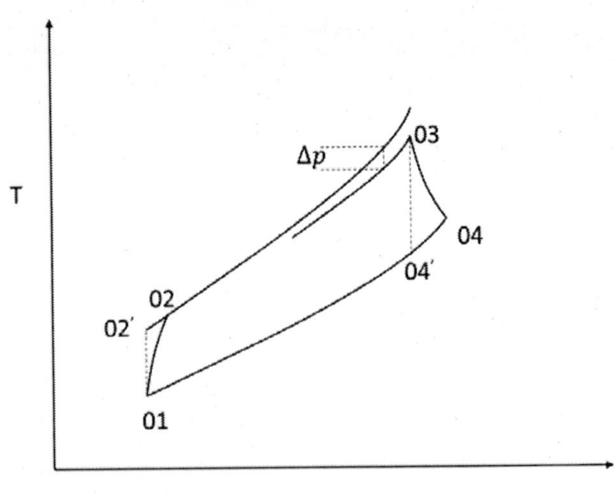

Figure 6.5: Actual Gas Turbine Cycle with Losses in Combustion Chamber

In the combustion chamber, there will be some pressure loss and therefore,

$$p_{03} = p_{02} - \Delta p$$

Temperature rise and fall during compression and expansion respectively,

$$(T_{02} - T_{01}) = \frac{T_{01}}{\eta_c}\left[r_c^{[(\gamma-1)/\gamma]_a} - 1\right]$$

$$(T_{03} - T_{04}) = \eta_T . T_{03}\left(1 - \frac{1}{r_t^{[(\gamma-1)/\gamma]_g}}\right)$$

As during compression process the fluid is air, suffix a and in turbine it is products of combustion i.e. hot gas, suffix g is used in the above relations.

Also, there will be some power loss during transmission of power from turbine to compressor, designated as mechanical efficiency, η_{mech},

Then net work output will be:

$$W_N = \eta_{mech}. C_{pg}. T_{03}. \eta_T\left[1 - \frac{1}{r_t^{[(\gamma-1)/\gamma]_g}}\right] - C_{pa}. T_{01}. \frac{1}{\eta_c}\left[r_c^{[(\gamma-1)/\gamma]_a} - 1\right]$$

Therefore the overall efficiency,

$$\eta = \frac{\text{Net work}}{\text{Heat supplied}} = \frac{\eta_{mech}. C_{pg}. (T_{03} - T_{04}) - C_{pa}. (T_{02} - T_{01})}{C_{pg}. (T_{03} - T_{02})}$$

6.6 COMBUSTION CHAMBER

The design of a combustion chamber is a complex process involving fluid dynamics, combustion, and mechanical design. The rapidly increasing use of computational fluid dynamics has a major impact on the design process. The most common fuel for a gas turbine is liquid petroleum distillate. There are two major considerations in design, one high combustion efficiency and the other is low pressure loss in the combustor. The typical values of combustion efficiency and pressure drop are more than 99.5% and 2–8% of the compressor delivery pressure. The other criteria are reliability, suitable temperature distribution in products of combustion without any hotspots at entry of the turbine and relighting in case of flame blow out during flight. The aircraft gas turbines also operate over a very wide range of air pressure and temperature within flight Mach number and altitude. Atmospheric conditions also change during climb and descent.

The factors affecting combustor design are:

1. The temperature of the gases leaving the combustion chamber should be low to suit the turbine blade material. With cooled blades, temperatures up to 1850 K are acceptable. With the improved turbine blade materials even higher temperatures may be possible.
2. The temperature distribution leaving the gases should be uniform to avoid local heating of blades in turbines. These are called hotspots.
3. Intake air velocity to the combustion chamber may be in the order of 30–60 m/s and stable operation is required for a wide range of air/fuel ratio. The air fuel ratio may vary from 60:1 to 120:1. With stoichiometric air fuel ratio of 15:1, high dilution is required to maintain the gas temperature suitable for turbine blades.
4. Stable combustion over a wide range of chamber pressure is required due to change in intake air parameters with altitude and forward speed. In the event of a flame blow out, relighting capability at high altitude is important.
5. There should be no smoke in the exhaust gases. Pollutants like NO_x, carbon monoxide and unburned hydrocarbon should also remain under control.

6.6.1 Combustor Types

Combustors are broadly divided into 3 categories, viz. can, can-annular and annular, shown in Figure 6.6. A can combustor consists of one or more cylindrical burners, each contained in a burner case. It was used in the early development period. The can-annular type of combustor consists of a series of cylindrical burners arranged within a common annulus, and hence the name

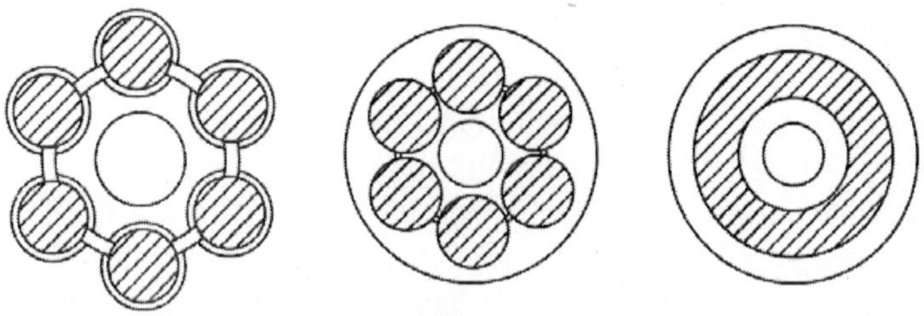

Figure 6.6: Types of Combustors

(*Source:* Lefebvre, Arthur H. and Ballal, Dilip R., Gas Turbine Combustion, 3rd Edition, CRC Press, 2010)

can-annular. Presently, these have been replaced by annular combustors in modern engines. In this system, a single burner having an annular cross section supplies hot gases to the turbine. The improved combustion zone, design, reduced linear surface area and shorter length have found it more suitable in new systems. Figure 6.7 gives the components of a typical annular combustor.

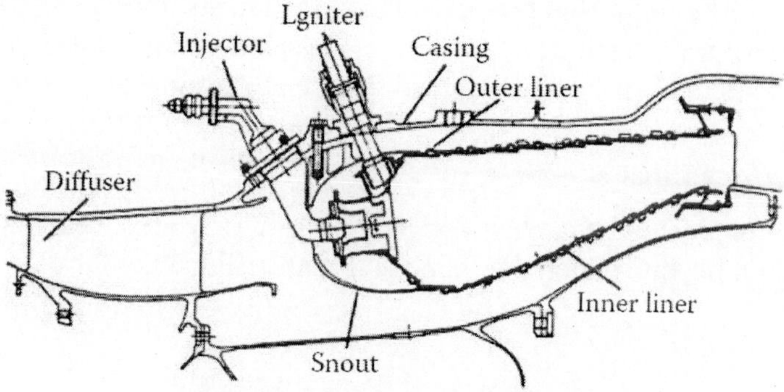

Figure 6.7: Annular Combustor

(*Source:* Lefebvre, Arthur H. and Ballal, Dilip R., Gas Turbine Combustion, 3rd Edition, CRC Press, 2010)

As the overall air-fuel ratio of a gas turbine combustor may range from 60:1 to 120:1, while the stoichiometric air-fuel ratio is 15:1, the air should be introduced in stages.

1. About 15–20% of the air is introduced in the primary zone to atomize the liquid fuel.
2. Some 30% air is introduced in the secondary zone to complete the fuel combustion.
3. The rest of the air is introduced in dilution zone to get the desired temperature and temperature distribution with no hot spots, to save the turbine blades.

6.6.2 Combustor Performance

The performance parameters are pressure loss, combustion efficiency, outlet temperature distribution, stability limits and combustion intensity.

(a) *Pressure Loss*

The overall stagnation pressure loss can be regulated as the sum of the fundamental loss, a small component which is the function of T_{02}/T_{01} and the frictional loss. The velocity of air leaving the compressor is high, say 150 m/s and a diffuser may be introduced before entry to the combustion chamber to reduce the velocity to 60 m/s. For aircraft gas turbines where space and weight are vital, the pressure loss should be preferably between 4 and 7%.

(b) *Combustion Efficiency*

The efficiency of a combustion process can be calculated from chemical analysis of the combustion products. Knowing air/fuel ratio and proportion of completely burnt constituents, the ratio of actual energy released from fuel and the theoretical energy available in fuel can be estimated.

(c) *Stability Limits*

For any combustion chamber there is both a rich and weak limit to the air/fuel ratio beyond which the flame is unstable (Figure 6.8).

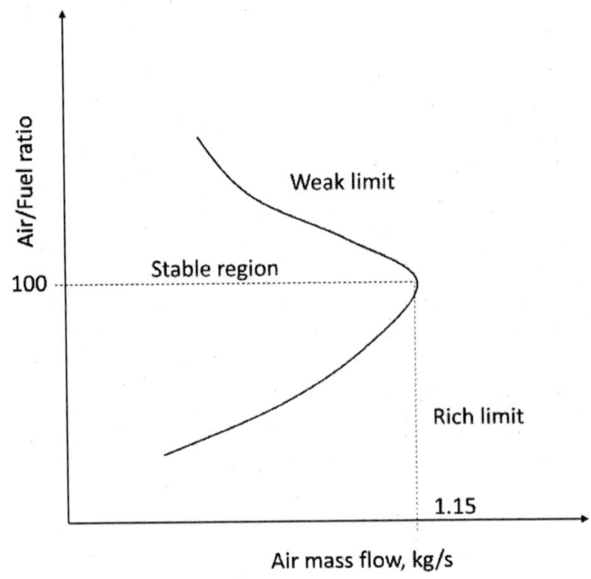

Figure 6.8: Typical Stability Curve

Usually, the limit is taken as the air/fuel ratio at which the flame blows out. Instability occurs before this limit is reached, which results in poor combustion and rough running. Allowances must be made for conditions when the engine is accelerated or decelerated. Most control systems have a built-in device which places an upper limit on the rate of change of fuel flow. The stability loop is a function of pressure in the combustion chamber. A decrease in pressure reduces the rate of chemical reaction and consequently narrows the stability limits.

(d) *Combustion Intensity*

The size of a combustion chamber is determined by the required rate of heat release from fuel. The nominal heat release rate can be estimated from air mass flow, fuel/air ratio and calorific value of the fuel. Larger the volume of combustion chamber, lower will be the pressure drop and higher combustion efficiency. It will also result in better outlet gas temperature distribution and improved stability characteristics. A term combustion intensity is introduced, and it is defined as:

$$\text{Combustion intensity} = \frac{\text{heat release rate}}{\text{combustion volume} \times \text{pressure}} \text{ kW/m}^3.\text{atm}$$

(e) *Exit Temperature Profile*

Exit gas temperature profile can be measured by means of a thermocouple or by gas analysis. A detailed profile of gas temperature, both radially and circumferentially, before entering the turbine, is prepared. A term, pattern factor or peak temperature is calculated from these data. The pattern factor (PF) is defined as:

$$PF = \frac{T_{0\,max} - T_{0\,avg}}{T_{0\,avg} - T_{0\,min}}$$

where, $T_{0\,\max}$ = maximum measured local exit gas temperature
$T_{0\,\text{avg}}$ = average of all temperatures at exit plane
$T_{0\,\min}$ = average of all temperatures at inlet plane

Acceptable values of pattern factor range from 0.25 to 0.45. By proper control of dilution air, the burner exit gas temperature profile is manipulated as per the turbine requirement.

6.6.3 Pollutant Formation and Control

Formation of NO_x depends on flame temperature and decreases for both rich and lean mixtures. When efforts are made to reduce NO_x, the formation of CO and unburned hydrocarbons HC, increases with decrease of flame temperature. Smoke formation is also important in the design of a combustion chamber and is not desirable. The rate of formation of NO_x varies exponentially with the flame temperature. NO_x formation also depends on fluid residence time in the combustor. NO_x emissions increase with the increase of combustor inlet temperature i.e., compressor delivery temperature. The pollutant problem can be tackled either by adjusting the conditions in the combustion chamber or by putting an exhaust clean up system.

6.6.4 Fuel Injection

The high-pressure fuel feed system is employed, and the liquid fuel is forced to a small orifice to form a conical spray of fine droplets in the primary zone. This is called fuel atomization. The fuel injection ports are tangential which impart a swirling action to the fuel flow. The atomized fuel in the air stream form droplets. The degree of atomization is expressed in terms of mean droplet diameter. The mean droplet diameter is in the range of 50–100 μm. The higher the supply pressure, the smaller will be the mean diameter. If the droplets are too small, they will not penetrate the airstream, and if too large the evaporation time will be too long. A compromise between the two aspects is the solution and fuel pressure should be enough for proper degree of atomization. A simple burner referred to as simplex will provide widely different atomizing properties over the fuel flow range. A duplex burner is more suitable. In this burner there are two fuel manifolds, each supplying independent orifices. It may be noted that the flow varies as the square root of the pressure. Figure 6.9 shows a schematic diagram of a swirl type atomizer. A good atomizer should fulfill the following requirements:

(a) Atomization must be affected over a wide range of fuel flows and pressure,
(b) design should be simple,
(c) should distribute the fuel uniformly throughout the air flows, and
(d) the weight and size should be low.

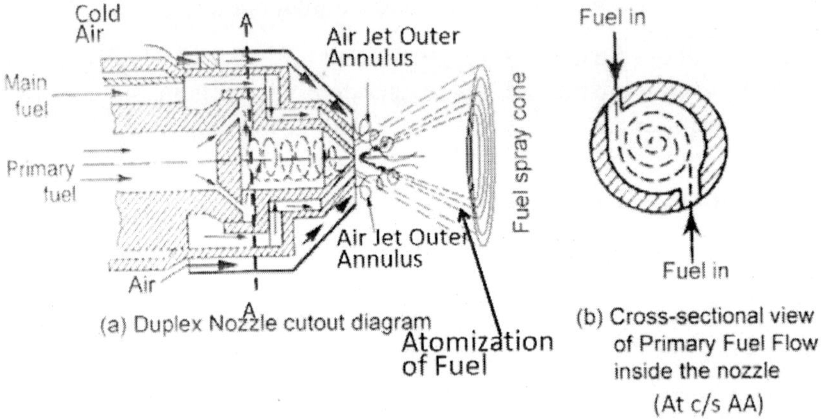

Figure 6.9: Swirl Type Atomizer

(*Source:* Roy, Bhaskar and Pradeep A.M., Jet Aircraft Propulsion, Lecture 26, IIT Bombay, NPTEL, 2012)

SUMMARY

In this chapter, we have studied the Brayton, the ideal gas turbine cycle, stagnation properties, compressor and turbine efficiencies, actual gas turbine cycle, combustion chamber and fuel injection systems. The efficiency of an ideal gas turbine depends on pressure ratio and the nature of the gas only. The typical values of combustion efficiency and pressure loss in a combustion chamber are 99.5% and 2–8% of the compressor delivery pressure. A simple burner referred to as simplex provides widely different atomizing properties over the fuel flow range. A duplex burner is more suitable. In this burner, there are two fuel manifolds, each supplying independent orifices.

EXAMPLES

Example 6.1

Evaluate the turbine pressure ratio for an ideal jet engine turbine having the following input data: Specific power output 450 kW,

Exhaust gas temperature leaving the turbine 850 K,

Take $\gamma = 1.4, C_p = 1.005$ kJ/(kg. K)

Solution

Refer Figure 6.1 (b)

$$W_T = 450\ kW$$

$$T_4 = 850\ K$$

$$W_T = C_p(T_3 - T_4)$$

$$450 = 1.005(T_3 - 850)$$

$$T_3 = \frac{450}{1.005} + 850$$

$$= 1297.76\ K$$

$$\frac{P_3}{P_4} = \left(\frac{T_3}{T_4}\right)^{\frac{\gamma}{(\gamma-1)}}$$

$$= \left(\frac{1297.76}{850}\right)^{\frac{1.4}{0.4}}$$

$$= 4.39.$$

Example 6.2

Evaluate the stagnation temperature at the exit of combustor having following input data:

Inlet air temperature = 400 K, Inlet air mass flow rate = 60 kg/s, Inlet fuel flow rate 1.5 kg/s, Heating value of fuel = 42800 kJ/kg, Burner efficiency = 100%, Take C_p for air = 1.005 kJ/(kg. K), C_p for products of combustion = 1.148 kJ/(kg. K)

Solution

Stagnation temperature at entry of combustor, $T_{02} = 400\ K$

Stagnation temperature at exit of combustor, $T_{03} = ?$

Burner efficiency, $\eta_b = 1$

The energy equation for combustion is as follows

$$\dot{m}_a . C_{pa} . T_{02} + \eta_b . \dot{m}_f . h_{PR} = (\dot{m}_a + \dot{m}_f) . C_{pg} . T_{03}$$

$$(60 \times 1.005 \times 10^3 \times 400) + (1 \times 1.5 \times 42800 \times 10^3) = (60 + 1.5) \times 1.148 \times 10^3 \times T_{03}$$

$$T_{03} = \frac{(60 \times 1.005 \times 10^3 \times 400) + (1 \times 1.5 \times 42800 \times 10^3)}{(60 + 1.5) \times 1.148 \times 10^3}$$

$$= 1250.95\ K.$$

Example 6.3

Evaluate the turbine efficiency for the given input data:

Pressure ratio across the turbine stage = 0.15, Turbine inlet total temperature = 1250 K, Turbine exit total temperature = 950 K, γ for products of combustion = 1.33.

Solution

Refer Figure 6.4.

$$\frac{T_{03}}{T_{04'}} = \left(\frac{p_{03}}{p_{04}}\right)^{(\gamma-1)/\gamma}$$

$$= \left(\frac{1}{0.15}\right)^{(1.33-1)/1.33}$$

$$\eta_t = \frac{T_{03} - T_{04}}{T_{03} - T_{04'}}$$

$$T_{03} - T_{04} = \eta_t \,.\, T_{03}\left(1 - \frac{T_{04'}}{T_{03}}\right)$$

$$1250 - 950 = \eta_t \times 1250 \times \left(1 - (0.15)^{0.33/1.33}\right)$$

$$1250 - 950 = \eta_t \times 1250 \times (1 - 0.624)$$

$$\eta_t = 0.639$$

$$= \mathbf{63.9\%}.$$

Example 6.4

Consider a gas turbine unit having the following input data

Air intake: pressure = 1 bar, temperature = 300 K, compressor output pressure = 8 bar, isentropic efficiency of compressor = 85%, combustion chamber exit gas temperature = 1100 K, turbine efficiency = 85%, heating value of fuel = 42,800 kJ/kg, air mass flow rate = 100 kg/min.

Calculate:

1. air/fuel ratio of turbine gases
2. final temperature of exhaust gases
3. net power produced
4. overall thermal efficiency

Neglect the effect of fuel in mass flow rate.

Assume, $C_{pa} = 1.005$ kJ/ kg. K , $\gamma_a = 1.4$
$C_{pg} = 1.148$ kJ/ kg. K , $\gamma_g = 1.33$

Solution

Refer Figure 6.10.

From the *T-s* diagram, shown in figure above,

$$\frac{T'_{02}}{T_{01}} = \left(\frac{p_{02}}{p_{01}}\right)^{\frac{(\gamma-1)}{\gamma}}$$

$$T'_{02} = 300 \times 8^{\frac{1.4-1}{1.4}}$$

$$= 300 \times 8^{0.286}$$

$$= 300 \times 1.81$$

$$= 543.7\ K$$

$$\eta_c = \frac{T'_{02} - T_{01}}{T_{02} - T_{01}}$$

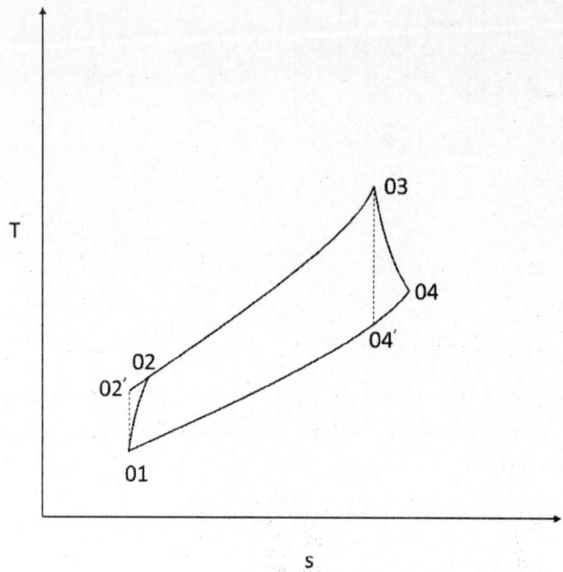

Figure 6.10: Real Gas Turbine Cycle

$$T_{02} = T_{01} + \frac{1}{\eta_c}(T'_{02} - T_{01}) = 300 + \frac{543 - 300}{0.85}$$

$$= 300 + \frac{243}{0.85} = 300 + 286.7$$

$$= 586.7\ K$$

$$\dot{m}_a . c_{pg} . (T_{03} - T_{02}) = \dot{m}_f . h_{PR}$$

$$\dot{m}_f = \frac{100 \times 1.148(1100 - 586.7)}{60 \times 42800}$$

$$= \mathbf{0.0229\ kg/s}$$

$$\text{Air/Fuel ratio} = \frac{\dot{m}_a}{\dot{m}_f} = \frac{100}{60 \times 0.0229} = 72.78$$

$$\frac{T'_{04}}{T_{03}} = \left(\frac{p_{04}}{p_{03}}\right)^{\frac{(\gamma-1)}{\gamma}}$$

$$T'_{04} = T_{03}\left(\frac{p_{04}}{p_{03}}\right)^{\frac{(\gamma-1)}{\gamma}}$$

$$= 1100\left(\frac{1}{8}\right)^{\frac{1.33-1}{1.33}}$$

$$= \frac{1100}{(8)^{0.248}}$$

$$= \frac{1100}{1.675}$$

$$= 656.7\ K$$

$$\eta_T = \frac{T_{03} - T_{04}}{T_{03} - T'_{04}}$$

$$T_{04} = T_{03} - \eta_T . (T_{03} - T'_{04})$$

$$= 1100 - 0.85(1100 - 656.7)$$

$$= \mathbf{723.4\ K}$$

Net power produced,

$$W_N = \dot{m}_a . C_{pg} . (T_{03} - T_{04}) - \dot{m}_a . C_{pa} . (T_{02} - T_{01})$$

$$= \left(\frac{100}{60} \times 1.148(1100 - 723.4)\right) - \left(\frac{100}{60} \times 1.005(586.7 - 300)\right)$$

$$= \mathbf{240.34\ kW}$$

Overall thermal efficiency,

$$\eta_T = \frac{W_N}{\dot{m}_f . h_{PR}}$$

$$= \frac{240.34}{0.0229 \times 42800}$$

$$= 0.245$$

$$= \mathbf{24.5\%}$$

Example 6.5

A simple turbojet unit operates with the following input data:

Maximum turbine inlet temperature = 1500 K, Pressure ratio = 6:1, Air mass flow = 30 kg/s, Isentropic efficiency of compressor = 85%, Isentropic efficiency of turbine = 90%, Propelling nozzle efficiency = 95%, Transmission efficiency = 98%, Combustion chamber pressure loss = 0.2 bar,

Assume, $C_{pa} = 1.005$ kJ/ kg. K, $\gamma = 1.4$ for air, $C_{pg} = 1.148$ kJ/kg. K, $\gamma = 1.33$ for products of combustion.

Calculate the design thrust produced.

Solution
Refer Figure 6.11

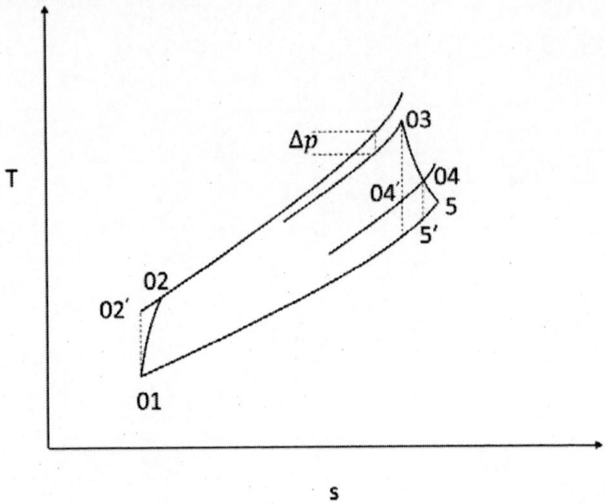

Figure 6.11: Real Turbojet Engine Cycle with Pressure Loss in Combustion Chamber

Ambient air condition, 1 bar, 300 K

01–02 : compressor
02–03 : combustion chamber
03–04 : turbine
04–05 : nozzle

$$\eta_c = \frac{T'_{02} - T_{01}}{T_{02} - T_{01}}$$

$$\frac{T'_{02}}{T_{01}} = (r_c)^{\frac{\gamma-1}{\gamma}}$$

$$(6)^{\left(\frac{1.4-1}{1.4}\right)} = 6^{0.286}$$

$$= 1.669$$

$$T'_{02} = 300 \times 1.669 = 500.8\ K$$

$$T_{02} = T_{01} + \frac{1}{\eta_c}(T'_{02} - T_{01})$$

$$= 300 + \frac{1}{0.85}(500.8 - 300)$$

$$= 300 + 236.24$$

$$= 536.24\ K.$$

Equate the work output of turbine to the work input to drive the compressor,

$$\eta_{tr}.C_{pg}(T_{03}-T_{04}) = C_{pa}(T_{02}-T_{01})$$

$$T_{03}-T_{04} = \frac{1.005\times 236.24}{0.98\times 1.148} = 211.03$$

$$T_{04} = T_{03} - 211.03 = 1500 - 211.03$$

$$= 1288.96$$

$$\eta_T = \frac{T_{03}-T_{04}}{T_{03}-T'_{04}}$$

$$T_{03}-T_{04} = \eta_T.T_{03}\left(1-\frac{T'_{04}}{T_{03}}\right)$$

$$= \eta_T.T_{03}\left(1-\frac{1}{(r_t)^{(r_g-1)/r_g}}\right)$$

where, r_t is turbine pressure ratio,

$$211.03 = 0.9\times 1500\left(1-\frac{1}{(r_t)^{0.248}}\right)$$

$$\frac{1}{(r_t)^{0.248}} = 1-0.1562$$

$$= 0.8436$$

$$r_t = \left(\frac{1}{0.8438}\right)^{11/0.248}$$

$$= (1.185)^{4.032}$$

$$= 1.983$$

$$p_{04} = \frac{p_{03}}{r_t} = \frac{p_{02}-\Delta p_{\text{comb.}}}{r_t}$$

$$= \frac{6-0.2}{r_t} = \frac{5.8}{r_t}$$

$$= \frac{5.8}{1.983} = 2.924 \text{ bar}$$

Consider nozzle,

$$\frac{T_{04}}{T'_5} = \left(\frac{p_{04}}{p_5}\right)^{\frac{(\gamma_g-1)}{\gamma_g}} = \left(\frac{2.924}{1}\right)^{(1.33-1)/1.33}$$

$$= (2.924)^{0.248} = 1.3$$

$$T'_5 = \frac{T_{04}}{1.3} = \frac{1288.96}{1.3}$$
$$= 991.5\ K$$

Nozzle efficiency,

$$\eta_n = \frac{T_{04} - T_5}{T_{04} - T'_5}$$
$$T_{04} - T_5 = 0.95(1288.96 - 991.5)$$
$$= 282.59\ K$$
$$T_5 = 1288.96 - 282.59 = 1006.37\ K$$

Jet velocity, V_j is same as V_5.

$$\frac{V_5^2}{2C_p} = \frac{V_j^2}{2C_p} = T_{04} - T_5$$
$$= 282.59$$
$$V_j = \sqrt{(2 \times 1.148 \times 1000 \times 282.59)}$$
$$= 805.5\ m/s$$

therefore, Thrust

$$= \dot{m}_a . V_j$$
$$= 30 \times 805.5\ m/s$$
$$= \mathbf{24165\ N}$$

QUESTIONS

1. Draw the p-V and T-s diagram of the Brayton cycle and briefly explain the working of a gas turbine system.
2. With the help of T-s diagram, give the expressions of isentropic efficiencies of compressor and turbine for a real jet engine.
3. Explain the functioning of a combustion chamber. Discuss the various arrangements that can be employed in aircraft engines.
4. Enumerate the various factors affecting the combustion chamber design.
5. Differentiate between simplex and duplex burners.

CHAPTER 7

Gas Dynamics

7.1 INTRODUCTION

In supersonic flow, compressibility and temperature become important. A new variable density and an extra equation, the equation of state, which relates pressure and density is considered. The other equations i.e., continuity, momentum and the first and second law of thermodynamics are also needed in the analysis of compressible fluid flow. The analysis is restricted to steady, one-dimensional flow of a perfect gas. Gas dynamics is also referred to as compressible fluid flow or high-speed aerodynamics.

7.2 PERFECT GAS RELATIONSHIPS

A perfect gas is defined as a fluid that has constant specific heats and that follows the equation,

$$p = \rho RT$$

where, p and T are the absolute pressure and absolute temperature respectively, ρ is the density, and R the gas constant.

For air, $\gamma = 1.4$ and $R = 287\ \text{J/kg. K}$

The specific heat at constant volume C_v is defined as,

$$C_v = \left(\frac{\partial u}{\partial T}\right)_v$$

where, u is internal energy per unit mass.

u is a function of temperature only for a perfect gas. The specific heat at constant pressure C_p is defined as,

$$C_p = \left(\frac{\partial h}{\partial T}\right)_p$$

where, h is the enthalpy per unit mass. Also,

$$h = u + p/\rho \text{ (by definition)}$$

$$h = u + RT$$

Therefore, h is also a function of temperature only.

For a perfect gas,

$$du = C_v\, dT$$

and,

$$dh = C_p\, dT$$

Now, differentiate equation,

$$h = u + RT$$

we get,

$$dh = du + RdT$$

$$\frac{dh}{dT} = \frac{du}{dT} + R$$

$$C_p = C_v + R$$

The specific heat ratio γ is defined as,

$$\gamma = C_p / C_v$$

from the above two equations,

$$\gamma\, C_v = C_v + R$$

$$(\gamma - 1)_{C_v} = R$$

$$C_v = \frac{R}{\gamma - 1}$$

and,

$$C_p = \frac{C_p}{\gamma} + R$$

$$C_p - \frac{C_p}{\gamma} = R$$

$$C_p \left(\frac{\gamma - 1}{\gamma}\right) = R$$

$$C_p = \frac{\gamma R}{\gamma - 1}$$

Entropy is defined by the relation,

$$ds = \frac{\delta_{q_{rev}}}{T}$$

where s is entropy and $\delta_{q_{rev}}$ is an incremental change of heat added reversibly to the system. Entropy is a measure of disorder.

Therefore, for a reversible process,

$$\delta q = T\,ds$$

In differential form,

$$dq = T\,ds$$

From first law of thermodynamics,

$$dq = du + pdv$$

also,

$$dq = dh - v\,dp$$

where, v is specific volume and

$$v = \frac{1}{\rho}$$

therefore,

$$Tds = dh - vdp$$

$$Tds = C_p\,dT - v\,dp$$

as,

$$pv = RT$$

$$v = \frac{RT}{p}$$

$$Tds = C_p\,dT - RT\frac{dp}{p}$$

$$ds = C_p\frac{dT}{T} - R\frac{dp}{p}$$

Integrate,

$$(s_2 - s_1) = C_p\int_1^2 \frac{dT}{T} - R\int_1^2 \frac{dp}{p}$$

$$(s_2 - s_1) = C_p\,ln\frac{T_2}{T_1} - R\,ln\frac{p_2}{p_1}$$

similarly,

$$Tds = du + pdv$$

$$Tds = C_v dT + \frac{RT}{v} dv$$

$$ds = C_v \frac{dT}{T} + R \frac{dv}{v}$$

Integrate,

$$(s_2 - s_1) = C_v \int_1^2 \frac{dT}{T} + R \int_1^2 \frac{dv}{v}$$

$$(s_2 - s_1) = C_v \, ln \frac{T_2}{T_1} + R \, ln \frac{v_2}{v_1}$$

also,

$$(s_2 - s_1) = C_p \, ln \frac{T_2}{T_1} - R \, ln \frac{p_2}{p_1}$$

For reversible adiabatic i.e., isentropic process

$$(s_2 - s_1) = 0$$

therefore,

$$0 = C_p \, ln \frac{T_2}{T_1} - R \, ln \frac{p_2}{p_1}$$

$$C_p \, ln \frac{T_2}{T_1} = R \, ln \frac{p_2}{p_1}$$

$$\frac{\gamma R}{(\gamma - 1)} . \ln\left(\frac{T_2}{T_1}\right) = R \, ln \left(\frac{p_2}{p_1}\right)$$

$$\ln\left(\frac{T_2}{T_1}\right) = \frac{(\gamma - 1)}{\gamma} ln \left(\frac{p_2}{p_1}\right)$$

$$= ln \left(\frac{p_2}{p_1}\right)^{\frac{(\gamma-1)}{\gamma}}$$

taking antilog,

$$\frac{T_2}{T_1} = \left(\frac{p_2}{p_1}\right)^{\frac{(\gamma-1)}{\gamma}}$$

also, when

$$(s_2 - s_1) = 0$$

$$0 = C_v \, ln \left(\frac{T_2}{T_1}\right) + R \, \ln\left(\frac{v_2}{v_1}\right)$$

$$-C_v\, ln\left(\frac{T_2}{T_1}\right) = R\ \ln\left(\frac{v_2}{v_1}\right)$$

$$-\frac{R}{(\gamma-1)}\, ln\left(\frac{T_2}{T_1}\right) = R\ln\left(\frac{v_2}{v_1}\right)$$

$$ln\left(\frac{T_2}{T_1}\right) = (\gamma-1)\, ln\left(\frac{v_1}{v_2}\right)$$

$$ln\left(\frac{T_2}{T_1}\right) = \ln\left(\frac{v_1}{v_2}\right)^{(\gamma-1)}$$

take antilog,

$$\frac{T_2}{T_1} = \left(\frac{v_1}{v_2}\right)^{(\gamma-1)}$$

as,

$$v = \frac{1}{\rho}$$

therefore,

$$\frac{T_2}{T_1} = \left(\frac{\rho_2}{\rho_1}\right)^{(\gamma-1)}$$

Finally, we have the following relationships for isentropic process,

$$\frac{T_2}{T_1} = \left(\frac{p_2}{p_1}\right)^{\frac{(\gamma-1)}{\gamma}} = \left(\frac{v_1}{v_2}\right)^{(\gamma-1)} = \left(\frac{\rho_2}{\rho_1}\right)^{(\gamma-1)}$$

7.3 SPEED OF SOUND WAVE AND MACH NUMBER

Consider steady fluid flow in a constant area, A, duct/channel (Figure 7.1). Let p, ρ, V be the pressure, density, and velocity in the upstream and $(p + dp)$, $(\rho + d\rho)$ and $(V + dV)$ be the same properties on downstream of the flow.

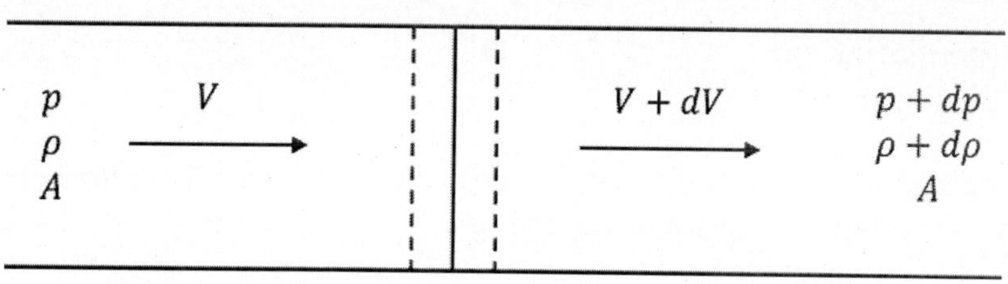

Figure 7.1: Steady Flow in a Constant Area Duct/Channel

Applying continuity equation,

$$\rho AV = \text{constant}$$

therefore,

$$\rho AV = (\rho + d\rho)A\,(V + dV)$$

$$\rho V = \rho V + \rho dV + V d\rho + d\rho . dV$$

$(d\rho . dV)$ being product of two very small quantities is neglected,

therefore,

$$\rho dV + V d\rho = 0$$

Now, apply momentum equation to the control volume within the dotted lines shown in Figure 7.1.

$$\text{Force} = \text{Rate of change of momentum}$$

$$= \text{mass} \times \text{change in velocity}$$

$$pA - (p + dp)A = \rho AV\,(V + dV - V)$$

$$-dp = \rho V dV$$

Substitute ρdV from equation,

$$-dp = -V.V\,d\rho$$

$$= -V^2 . d\rho$$

$$dp = V^2 d\rho$$

$$V^2 = dp/d\rho$$

Therefore, a small disturbance or sudden change in conditions in steady flow can occur only when a particular velocity, $V = \sqrt{(dp/d\rho)}$ exists in the channel. If a uniform velocity, $V = \sqrt{(dp/d\rho)}$ is assumed to the left of the control volume (Figure 7.1), the continuity and momentum equations are valid, and the small disturbance is propagated through a fluid at rest. This is called the speed of sound, a in the medium. Large disturbances may travel faster than the speed of sound, e.g., bomb explosion.

The speed of sound,

$$a = \sqrt{(dp/d\rho)}$$

The bulk modulus, K, is defined as the ratio of the change in pressure dp to change in volume, $d\forall$ per unit volume ($\forall$).

$$K = -\frac{dp}{d\forall/\forall}$$

The negative sign indicates that when pressure increases, the volume will decrease. $\forall$ is the volume of the fluid subjected to pressure change dp. Also,

$$\frac{d\forall}{\forall} = \frac{dv}{v} = -\frac{d\rho}{\rho}$$

therefore,

$$K = \frac{-dp}{-d\rho/\rho}$$

$$K = \rho\frac{dp}{d\rho}$$

$$K = \rho a^2$$

$$a = \sqrt{K/\rho}$$

for a perfect gas,

$$p = \rho RT$$

for isentropic flow,

$$p\rho^{-\gamma} = \text{constant}$$

take log and differentiate,

$$\frac{dp}{p} - \gamma\frac{d\rho}{\rho} = 0$$

$$\frac{dp}{p} = \gamma\frac{d\rho}{\rho}$$

$$\frac{dp}{d\rho} = \gamma\frac{p}{\rho}$$

$$a^2 = \gamma RT$$

$$a = \sqrt{\gamma RT}$$

This shows that the speed of sound in a perfect gas is a function of absolute temperature only. In isothermal flow the speed of sound remains constant.

The Mach number is defined as the ratio of velocity of fluid to the local velocity of sound in the medium.

$$M = \frac{V}{a}$$

$$M^2 = \frac{V^2}{a^2}$$

This may be interpreted as the ratio of kinetic energy of the fluid to its thermal energy, as the kinetic energy is proportional to V^2 and thermal energy is proportional to T. The Mach number is a measure of the importance of compressibility. At $M = 0.3$ and above, the compressibility effects are predominant. Below this, the fluid may be considered incompressible. i.e., density remains constant. Flows with Mach number more than unity are referred to as supersonic flows. For convenience, the classification of the regimes is as follows:

1. Subsonic flow, $0 < M < 0.8$
2. Transonic flow, $0.8 < M < 1.2$
3. Supersonic flow, $1.2 < M < 5$
4. Hypersonic flow, $M > 5$

In transonic flow, the behavior may be subsonic in some portions and supersonic over other portions. There is a possibility of shock on the body. It is the mixed nature of the flow which makes the analysis more complicated.

7.4 MACH CONE

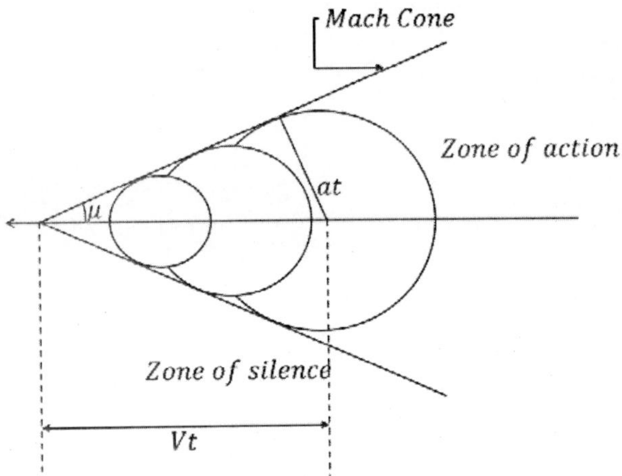

Figure 7.2: Propagation of a Supersonic Disturbance ($V > a$)

It is known that a small disturbance is observed throughout the field and the wave travels at local velocity of sound relative to the medium. The disturbance wave reaches a stationary observer before the source of disturbance reaches him in subsonic flow. In supersonic flow, it takes more time for an observer to perceive the pressure disturbance, after the source has passed him. This is the basic difference between subsonic and supersonic flow. In supersonic flow, a well-defined conical zone is formed with the object located at the nose of the cone and the disturbance is confined only to the field inside the cone. The region inside the cone is known as a zone of action and the region outside the cone is known as a zone of silence. The lines of which the pressure disturbance is concentrated, and which generate the cone are called Mach waves or Mach lines. The angle between the Mach line and the direction of motion of the body is called the Mach angle μ.

$$\sin\mu = \frac{\text{at}}{Vt} = \frac{a}{V}$$

$$\sin\mu = \frac{1}{M}$$

where, t is time.

7.5 ISENTROPIC FLOW

A frictionless adiabatic i.e., isentropic flow is an ideal flow which can never be achieved in the flow of real gases. Consider the one-dimensional steady flow of a perfect gas through a converging and converging-diverging duct/nozzle.

Euler's equation in differential form, neglecting elevation changes is,

$$V\,dV + \frac{dp}{\rho} = 0$$

and the continuity equation,

$$\rho AV = \text{constant}$$

Differentiate and divide by ρAV

$$\frac{d\rho}{\rho} + \frac{dA}{A} + \frac{dV}{V} = 0$$

$$\frac{d\rho}{\rho} = -\left(\frac{dA}{A} + \frac{dV}{V}\right)$$

by definition,

$$a^2 = \frac{dp}{d\rho}$$

therefore, from Euler's equation,

$$VdV + a^2\frac{d\rho}{\rho} = 0$$

eliminate $\frac{d\rho}{\rho}$ using continuity equation,

$$VdV + a^2\left(-\frac{dA}{A} - \frac{dV}{V}\right) = 0$$

$$VdV - \frac{a^2}{V}dV = a^2\frac{dA}{A}$$

$$\frac{dV}{V}(V^2 - a^2) = a^2\frac{dA}{A}$$

$$\frac{dA}{dV} = \frac{A}{V}\left(\frac{V^2}{a^2} - 1\right)$$

$$\frac{dA}{dV} = \frac{A}{V}(M^2 - 1)$$

$$\frac{dA}{dV} = -\frac{A}{V}(1 - M^2)$$

This shows that for subsonic flow (M<1), $\frac{dA}{dV}$ is always negative, i.e., the channel area should decrease for increasing velocity. For M=1, $\frac{dA}{dV} = 0$, the velocity keeps increasing until the minimum section or throat is reached and that is the only section where sonic flow may occur. Also, for supersonic flow (M>1) $\frac{dA}{dV}$ is positive and the area must increase for an increase in velocity. Hence, to produce a supersonic steady flow from a fluid at rest in a reservoir, it must pass flow through a converging duct and then a diverging duct.

For an adiabatic process, the energy equation is

$$h_1 + \frac{V_1^2}{2} = h_2 + \frac{V_2^2}{2}$$

For state 1, let stagnation enthalpy h_0 and velocity $V_1 = 0$

$$h_0 = h + \frac{V^2}{2}$$

where h and V are enthalpy and velocity in any other location.

$$C_pT_0 = C_pT + \frac{V^2}{2}$$

$$T_0 = T + \frac{1}{C_p}\cdot\frac{V^2}{2}$$

$$T_0 = T + \frac{\gamma - 1}{\gamma R}\cdot\frac{V^2}{2}$$

$$\frac{T_0}{T} = 1 + \frac{\gamma - 1}{\gamma RT} \cdot \frac{V^2}{2}$$

as, $a^2 = \gamma RT$

$$\frac{T_0}{T} = 1 + \frac{\gamma - 1}{a^2} \cdot \frac{V^2}{2}$$

$$\frac{T_0}{T} = 1 + \frac{\gamma - 1}{2} M^2$$

from isentropic relations,

$$\frac{p_0}{p} = \left(\frac{T_0}{T}\right)^{\gamma/(\gamma-1)}$$

$$\frac{p_0}{p} = \left(1 + \frac{\gamma - 1}{2} M^2\right)^{\gamma/(\gamma-1)}$$

and,

$$\frac{\rho_0}{\rho} = \left(\frac{T_0}{T}\right)^{1/(\gamma-1)}$$

$$\frac{\rho_0}{\rho} = \left(1 + \frac{\gamma - 1}{2} M^2\right)^{1/(\gamma-1)}$$

Conditions at throat can be calculated by putting $M = 1$. Sonic conditions are marked with asterisk.

$$M = 1, \qquad a^* = V^* = \sqrt{\gamma RT^*}$$

$$\frac{T_0}{T^*} = 1 + \frac{\gamma - 1}{2}$$

$$\frac{T_0}{T^*} = \frac{\gamma + 1}{2}$$

$$\frac{T^*}{T_0} = \frac{2}{\gamma + 1}$$

for air, $\gamma = 1.4$

$$\frac{T^*}{T_0} = 0.833$$

$$\frac{p^*}{p_0} = \left(\frac{2}{\gamma + 1}\right)^{\gamma/(\gamma-1)}$$

$$= 0.528$$

$$\frac{\rho^*}{\rho_0} = \left(\frac{2}{\gamma+1}\right)^{1/(\gamma-1)} = 0.634$$

From these relations, it can be inferred that the absolute temperature drops by about 17% from reservoir to throat, the critical pressure is 52.8% of reservoir pressure and density is reduced by about 37%.

The variation of area with Mach number for the critical case is calculated by using the continuity equation,

$$\rho AV = \rho^* A^* V^*$$

A^* is minimum at throat, Then

$$\frac{A}{A^*} = \frac{\rho^*}{\rho} \cdot \frac{V^*}{V}$$

now,

$$V^* = a^* = \sqrt{\gamma R T^*}$$

and,

$$V = M.a$$

$$V = M\sqrt{\gamma R T}$$

therefore,

$$\frac{V^*}{V} = \frac{1}{M}\sqrt{\frac{T^*}{T}}$$

$$= \frac{1}{M}\sqrt{\frac{T^*}{T_0}} \cdot \sqrt{\frac{T_0}{T}}$$

$$= \frac{1}{M}\sqrt{\frac{2}{(\gamma+1)}} \cdot \sqrt{\left(1+\frac{\gamma-1}{2}M^2\right)}$$

$$\frac{V^*}{V} = \frac{1}{M}\left[\frac{2}{(\gamma+1)} \cdot \left(1+\frac{\gamma-1}{2}M^2\right)\right]^{1/2}$$

and,

$$\frac{\rho^*}{\rho} = \frac{\rho^*}{\rho_0} \cdot \frac{\rho_0}{\rho}$$

$$\frac{\rho^*}{\rho} = \left(\frac{2}{\gamma+1}\right)^{1/(\gamma-1)} \cdot \left(1+\frac{\gamma-1}{2}M^2\right)^{1/(\gamma-1)}$$

substituting these values,

$$\frac{A}{A^*} = \frac{1}{M}\left[\frac{2}{(\gamma+1)}\left(1+\frac{\gamma-1}{2}M^2\right)\right]^{\left(\frac{1}{(\gamma-1)}+\frac{1}{2}\right)}$$

$$\frac{A}{A^*} = \frac{1}{M}\left[\frac{2}{(\gamma+1)}\left(1+\frac{\gamma-1}{2}M^2\right)^{\frac{(\gamma+1)}{2(\gamma-1)}}\right]$$

This is Area-Mach number relation for a convergent divergent nozzle. It may be noted that $A/_{A^*}$ will never be less than unity, and for any value greater than unity there will be two values of Mach number, one less than one and other more than one. For air $\gamma = 1.4$,

$$\frac{A}{A^*} = \frac{1}{M}\left(\frac{5+M^2}{6}\right)^3$$

At throat the mass flow rate of fluid will be maximum. Therefore,

$$\dot{m}_{max} = \rho^* A^* V^*$$

$$= \rho_0 . \left(\frac{2}{\gamma+1}\right)^{1/(\gamma-1)} . A^*\sqrt{\gamma R T^*}$$

$$= \rho_0 . \left(\frac{2}{\gamma+1}\right)^{1/(\gamma-1)} . A^*\sqrt{\frac{\gamma R 2 T_0}{(\gamma+1)}}$$

$$\left(as, \rho_0 = p_0/_{RT_0}\right)$$

$$\dot{m}_{max} = \frac{A^* p_0}{RT_0}\left(\frac{2}{(\gamma+1)}\right)^{1/(\gamma-1)} . \sqrt{\frac{\gamma R 2 T_0}{(\gamma+1)}}$$

$$\dot{m}_{max} = \frac{A^* p_0}{\sqrt{T_0}}\sqrt{\frac{\gamma}{R} . \left(\frac{2}{(\gamma+1)}\right)^{\left(\frac{\gamma+1}{\gamma-1}\right)}}$$

for air $\gamma = 1.4$,

$$\dot{m}_{max} = 0.686\frac{A^* p_0}{\sqrt{RT_0}}$$

For adiabatic flow, the energy equation is

$$h_2 + \frac{V_2^2}{2} = h_1 + \frac{V_1^2}{2}$$

$$C_pT_2 + \frac{V_2^2}{2} = C_pT_1 + \frac{V_1^2}{2}$$

$$\frac{\gamma RT_2}{\gamma - 1} + \frac{V_2^2}{2} = \frac{\gamma RT_1}{\gamma - 1} + \frac{V_1^2}{2}$$

from equation of state, $p_1 = \rho RT_1$, $p_2 = \rho RT_2$

$$\frac{\gamma}{\gamma - 1} \cdot \frac{p_2}{\rho_2} + \frac{V_2^2}{2} = \frac{\gamma}{\gamma - 1} \cdot \frac{p_1}{\rho_1} + \frac{V_1^2}{2}$$

Assign state 2 as general any state and state 1, the stagnation conditions, $V_1 = 0$, $p_1 = p_0$ and $\rho_1 = \rho_0$.

$$\frac{\gamma}{\gamma - 1} \cdot \frac{p}{\rho} + \frac{V^2}{2} = \frac{\gamma}{\gamma - 1} \cdot \frac{p_0}{\rho_0}$$

also, $a = \sqrt{\gamma RT}$;

$$a^* = \sqrt{\gamma RT^*}, V^* = a^*$$

$$\frac{a^2}{\gamma - 1} + \frac{V^2}{2} = \frac{{a^*}^2}{\gamma - 1} + \frac{{a^*}^2}{2}$$

$$= \frac{{a^*}^2(\gamma + 1)}{2(\gamma - 1)}$$

divide by V^2,

$$\frac{(a/V)^2}{(\gamma - 1)} + \frac{1}{2} = \frac{(\gamma + 1)}{2(\gamma - 1)} \cdot \left(\frac{a^*}{V}\right)^2$$

$$\frac{(1/M)^2}{(\gamma - 1)} + \frac{1}{2} = \frac{(\gamma + 1)}{2(\gamma - 1)} \cdot \left(\frac{1}{M^*}\right)^2$$

$$\left(\frac{1}{M}\right)^2 = (\gamma - 1)\left[\frac{(\gamma + 1)}{2(\gamma - 1)} \cdot \left(\frac{1}{M^{*2}} - \frac{1}{2}\right)\right]$$

$$= \frac{(\gamma + 1)}{2M^{*2}} - \frac{(\gamma - 1)}{2}$$

$$M^2 = \frac{1}{\left[\frac{(\gamma+1)}{2M^{*2}} - \frac{(\gamma-1)}{2}\right]}$$

$$M^2 = \frac{2}{\left[\frac{(\gamma+1)}{M^{*2}} - (\gamma-1)\right]}$$

similarly,

$$\frac{\left(1/_{M}\right)^{2}}{(\gamma-1)}+1/_{2}=\frac{\gamma+1}{2(\gamma-1)}\left(\frac{1}{M^{*2}}\right)$$

$$\frac{1}{M^{*2}}=\frac{2(\gamma-1)}{(\gamma+1)}\left[\frac{1}{(\gamma-1)M^{2}}+\frac{1}{2}\right]$$

$$=\left[\frac{2}{(\gamma+1)M^{2}}+\frac{(\gamma-1)}{(\gamma+1)}\right]$$

$$=\left[\frac{2+(\gamma-1)M^{2}}{(\gamma+1)M^{2}}\right]$$

$$M^{*^{2}}=\frac{\frac{\gamma+1}{2}M^{2}}{\left(1+\frac{\gamma-1}{2}M^{2}\right)}$$

7.6 NOZZLE FLOW

A nozzle is a passage to transform pressure energy into kinetic energy. A convergent-divergent nozzle (C-D) is used to generate supersonic flow. A convergent-divergent nozzle is also called De Laval Nozzle, named after the person who used it first in his steam turbine research in nineteenth century (Figure 7.3). At the throat of the nozzle, the flow is sonic.

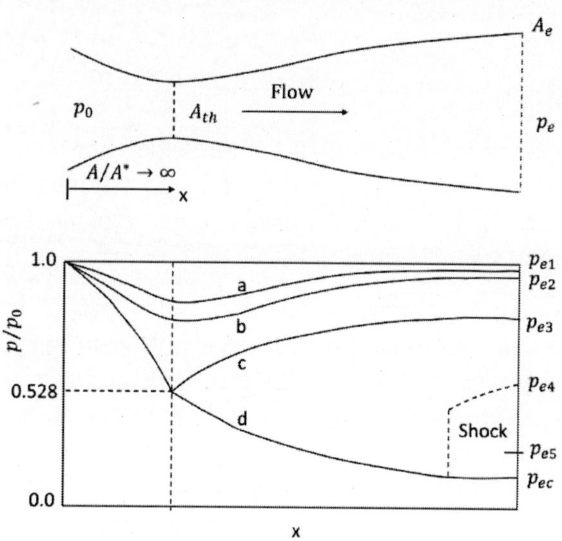

Figure 7.3: Flow in a Convergent-Divergent Nozzle

(*Source:* Rathakrishnan, E.., Gas Dynamics, PHI Learning Private Limited, New Delhi, Second Edition, 2009)

From Area Mach number relation, it may be observed that $M = f(A/A^*)$, where f is a function. For any value of A/A^*, we get two values of M, one for subsonic flow and the other for supersonic flow. As A/A^* less than one is not possible, and is always more than unity, therefore, the pressure, temperature and density decrease continuously throughout the nozzle. The exit pressure, density, and temperature ratios p_e/p_0, ρ_e/ρ_0 and T_e/T_0 depend only on the exit area ratio A_e/A^*. The properties of the gas in reservoir are p_0, ρ_0 and T_0. To accelerate the flow of air, a favourable pressure gradient needs to be created across the nozzle. The pressure at exit should be lower than the inlet pressure, i.e., p_e/p_0<1. Supersonic velocity can be reached only when p_e/p_0<0.528, and the Mach number at the throat will be unity and the flow is designated as 'choked flow' ($M^* = 1$ at the throat). Any further decrease in pressure will result in supersonic flow in the divergent portion of the nozzle.

A variety of flow fields can be generated by governing the back pressure downstream of the nozzle exit. The mass flow of air is maximum when the flow is choked and any further decrease in pressure at exit will not result in increase of mass flow of air through the nozzle. The pressure of choked flow is also called the back pressure. The critical pressure is the exit pressure required for shock free operation.

When the exit pressure, $p_e < p_b$, the nozzle is said to be over expanded and when $p_e > p_b$, the nozzle is said to be under expanded. Where, p_b is back pressure.

7.7 SHOCK WAVES

In one-dimensional flow, only normal compression shock wave can occur. Shock may be defined as a compression front in a supersonic flow field. Figures 7.4 (a) and (b) show the streamlines around a blunt faced cylinder in subsonic flow and supersonic flow.

Due to shock formation, abrupt change in fluid properties takes place. In subsonic flow, the molecules get information about the presence of the body, and they adjust themselves. But when the incoming stream is supersonic, the molecules travel faster than the signals and they do not get the information about the presence of body and the reflected signals from the face of the cylinder tend to accumulate at a short distance ahead of the body and form a thin compression front called shock wave as shown in Figure 7.4 (b). The flow is subsonic after a normal shock. The thickness of the shock is comparable to the mean free path of the gas molecules in the flow field.

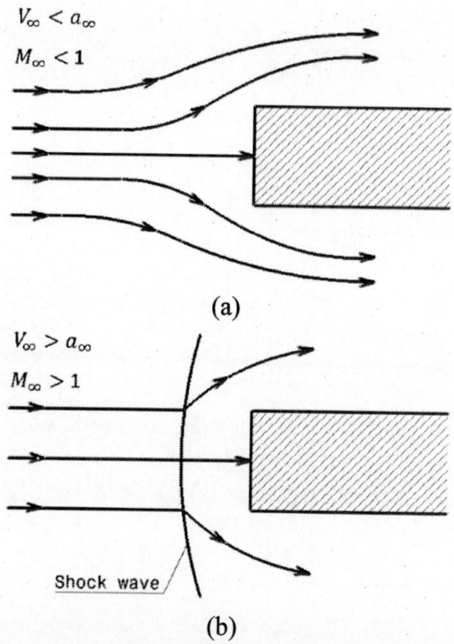

Figure 7.4: Streamlines Around a Blunt Face Cylinder in (a) Subsonic Flow and (b) Supersonic Flow

(*Source:* Rathakrishnan, E., Gas Dynamics, PHI Learning Private Limited, New Delhi, Second Edition, 2009)

Consider an adiabatic flow through a constant area duct as shown in Figure. 7.5. Let sections (1) and (2) be sufficiently away from the region.

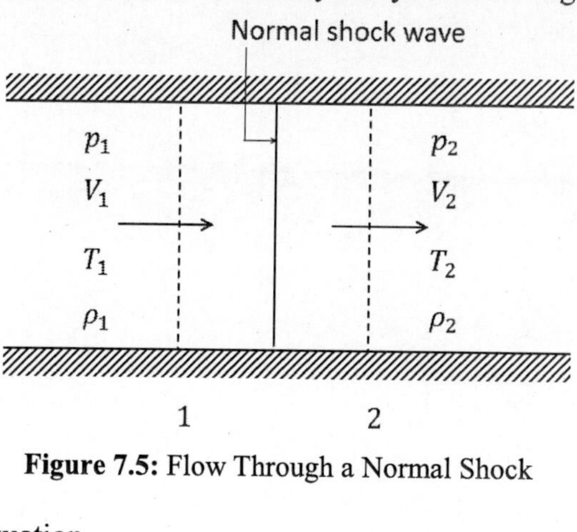

Figure 7.5: Flow Through a Normal Shock

The continuity equation

$$\rho_1 V_1 = \rho_2 V_2$$

and the momentum equation,

$$(p_1 - p_2)A = m(V_2 - V_1)$$
$$= (mV_2 - mV_1)$$
$$= (\rho_2 V_2 V_2 - \rho_1 V_1 V_1)A$$
$$(p_1 - p_2) = \rho_2 {V_2}^2 - \rho_1 {V_1}^2$$
$$p_1 + \rho_1 V_1^2 = p_2 + \rho_2 V_2^2$$

and the energy equation,

$$h_1 + \frac{V_1^2}{2} = h_2 + \frac{V_2^2}{2}$$

The control sections (1) and (2) may be brought close. Such a front or discontinuity across which there is sudden change in fluid properties is called a shock wave. When no heat is added or taken away, the flow across the shock wave is adiabatic.

For a calorically perfect gas,

$$p = \rho RT$$

and,

$$h = C_p T$$

Divide momentum equation by continuity equation,

$$\frac{p_1}{\rho_1 V_1} - \frac{p_2}{\rho_2 V_2} = V_2 - V_1$$

from equation of state,

$$\frac{RT_1}{V_1} - \frac{RT_2}{V_2} = V_2 - V_1$$

also, $a_1 = \sqrt{\gamma RT_1}$ and $a_2 = \sqrt{\gamma RT_2}$

$$\frac{a_1^2}{\gamma V_1} - \frac{a_2^2}{\gamma V_2} = V_2 - V_1$$

from energy equation,

$$C_p T_1 + \frac{V_1^2}{2} = C_p T_2 + \frac{V_2^2}{2}$$

$$\frac{\gamma RT_1}{\gamma - 1} + \frac{V_1^2}{2} = \frac{\gamma RT_2}{\gamma - 1} + \frac{V_2^2}{2}$$

$$\frac{{a_1}^2}{\gamma - 1} + \frac{V_1^2}{2} = \frac{{a_2}^2}{\gamma - 1} + \frac{V_2^2}{2}$$

Let the section 2 be throat

$$V_2 = a_2 = a^*$$

$$\frac{{a_1}^2}{\gamma - 1} + \frac{V_1^2}{2} = \frac{a^{*2}}{\gamma - 1} + \frac{a^{*2}}{2}$$

$$= \frac{(\gamma + 1)a^{*2}}{2(\gamma - 1)}$$

$${a_1}^2 = \frac{(\gamma + 1)a^{*2}}{2} - \frac{(\gamma - 1)}{2} V_1^2$$

similarly,

$${a_2}^2 = \frac{(\gamma + 1)\, a^{*2}}{2} - \frac{(\gamma - 1)}{2} V_2^2$$

substitute a_1 and a_2 in the equation

$$\frac{a_1^2}{\gamma V_1} - \frac{a_2^2}{\gamma V_2} = V_2 - V_1$$

$$\frac{1}{\gamma V_1}\left[\frac{(\gamma + 1)a^{*2}}{2} - \frac{(\gamma - 1)}{2} V_1^2\right] - \left[\frac{(\gamma + 1)\, a^{*2}}{2} - \frac{(\gamma - 1)}{2} V_2^2\right]\frac{1}{\gamma V_2}$$

$$= V_2 - V_1$$

$$\frac{(\gamma + 1)a^{*2}}{2\gamma V_1} - \frac{(\gamma + 1)a^{*2}}{2\gamma V_2} - \frac{\gamma - 1}{2\gamma} V_1 + \frac{\gamma - 1}{2\gamma} V_2 = V_2 - V_1$$

$$\frac{(\gamma + 1)a^{*2}}{2\gamma}\left[\frac{1}{V_1} - \frac{1}{V_2}\right] - \left(\frac{\gamma - 1}{2\gamma}\right)(V_1 - V_2)$$

$$= V_2 - V_1$$

$$\frac{(\gamma + 1)a^{*2}}{2\gamma} . \frac{(V_2 - V_1)}{V_1 V_2} + \frac{(\gamma - 1)}{2\gamma}(V_2 - V_1) = (V_2 - V_1)$$

divide by $(V_2 - V_1)$

$$\frac{(\gamma + 1)a^{*2}}{2\gamma V_1 V_2} + \frac{(\gamma - 1)}{2\gamma} = 1$$

$$\frac{(\gamma + 1)a^{*2}}{2\gamma V_1 V_2} = 1 - \frac{(\gamma - 1)}{2\gamma}$$

$$= \frac{2\gamma - \gamma + 1}{2\gamma}$$

$$= \frac{(\gamma + 1)}{2\gamma}$$

$$\frac{a^{*2}}{V_1 V_2} = 1$$

$$a^{*2} = V_1 V_2$$

This is called Prandtl relation.

as,

$$M_1^* = \frac{V_1}{a^*} \text{ and } M_2^* = \frac{V_2}{a^*}$$

$$M_1^* . M_2^* = 1$$

$$M_2^* = \frac{1}{M_1^*}$$

This shows that velocity changes across a normal shock should be from supersonic to subsonic or vice-versa. Flow from subsonic to supersonic is not possible because the entropy change will then be negative which is not thermodynamically possible. Therefore, the normal shock is formed only when the flow is supersonic and after the shock the flow is always subsonic. Earlier, it has been shown that,

$$M^{*2} = \frac{(\gamma+1)M^2}{(\gamma+1)M^2+2}$$

therefore,

$$M_1^{*2} = \frac{(\gamma+1)M_1^2}{(\gamma-1)M_1^2+2}$$

and

$$M_2^{*2} = \frac{(\gamma+1)M_2^2}{(\gamma-1)M_2^2+2}$$

substitute in the equation,

$$M_2^{*2} = \frac{1}{M_1^{*2}}$$

$$M_2^2 = \frac{\left(1+\frac{\gamma-1}{2}M_1^2\right)}{\left(\gamma M_1^2 - \frac{\gamma-1}{2}\right)}$$

When $M_1 = 1, M_2 = 1$

This is defined as Mach wave, an infinitely weak normal shock. As M_1 increases the normal shock will become stronger and M_2 will be progressively less than 1.

under limit as $M_1 \to \infty$

$$M_2^2 = \frac{\frac{(\gamma-1)}{2} M_1^2}{\gamma M_1^2}$$

$$= \frac{(\gamma - 1)}{2\gamma}$$

$$M_2 \to \sqrt{\frac{(\gamma - 1)}{2\gamma}}$$

for air $\gamma = 1.4$, $M_2 = 0.378$

now,

$$\frac{V_1}{V_2} = \frac{V_1.V_1}{V_2.V_1} = \frac{V_1^2}{V_1.V_2} = \frac{V_1^2}{a^{*2}} = M_1^{*2}$$

$$M_1^{*2} = \frac{(\gamma + 1)M_1^2}{[(\gamma - 1)M_1^2 + 2]}$$

from continuity equation,

$$\frac{\rho_2}{\rho_1} = \frac{V_1}{V_2}$$

$$= \frac{(\gamma + 1)M_1^2}{[(\gamma - 1)M_1^2 + 2]}$$

from momentum equation,

$$p_2 - p_1 = \rho_1 V_1^2 - \rho_2 V_2^2$$

divide by p_1,

$$\frac{p_2 - p_1}{p_1} = \frac{\rho_1 V_1^2 - \rho_2 V_2^2}{p_1}$$

$$= \frac{\rho_1 V_1^2}{p_1}\left(1 - \frac{\rho_2 V_2^2}{\rho_1 V_1^2}\right)$$

$$= \frac{\rho_1 V_1^2}{p_1}\left(1 - \frac{V_2}{V_1}\right)$$

$$(as, \rho_1 V_1 = \rho_2 V_2)$$

also,

$$\frac{p_1}{\rho_1} = RT_1 = \frac{a_1^2}{\gamma}$$

$$\frac{p_2 - p_1}{p_1} = \gamma M_1^2 \left(1 - \frac{2 + (\gamma - 1)M_1^2}{(\gamma + 1)M_1^2}\right)$$

$$\frac{p_2}{p_1} = 1 + \frac{2\gamma}{(\gamma + 1)}(M_1^2 - 1)$$

The ratio $\frac{p_2 - p_1}{p_1}$ is called shock strength.

From equation of state,

$$\frac{T_2}{T_1} = \left(\frac{p_2}{p_1}\right)\left(\frac{\rho_1}{\rho_2}\right)$$

The entropy change can be calculated using the equation,

$$(s_2 - s_1) = C_p \, ln\frac{T_2}{T_1} - R \, ln\frac{p_2}{p_1}$$

for air $\gamma = 1.4$

under limit when $M_1 \to \infty$

$$\lim_{M_1 \to \infty} M_2 = \sqrt{\frac{\gamma - 1}{2\gamma}}$$

$$= 0.378$$

$$\lim_{M_1 \to \infty} \frac{\rho_2}{\rho_1} = \frac{\gamma + 1}{\gamma - 1} = 6$$

$$\lim_{M_1 \to \infty} \frac{p_2}{p_1} = \infty$$

$$\lim_{M_1 \to \infty} \frac{T_2}{T_1} = \infty$$

7.7.1 Oblique Shock

In the previous section, the normal shock, which is a compression wave normal to the flow direction has been studied. In a wide variety of situations, a compression wave inclined at an angle to the flow occurs. This is called an oblique shock. Oblique shocks normally occur when a supersonic flow is turned into itself and when the flow is turned away from itself, it is designated as expansion fan. These are shown in Figure 7.6 (a) and (b). Angle θ is referred to as flow deflection angle. In compression corner, the streamlines are deflected by angle, θ and the flow is uniform parallel flow downstream of the shock. After the shock wave, the Mach number decreases, and the pressure, density, and temperature increase. In an expansion corner, Mach number increases, and the pressure, density, and temperature decrease.

Oblique shock and expansion waves prevail in two- and three-dimensional supersonic flows, while the normal shock waves are one dimensional.

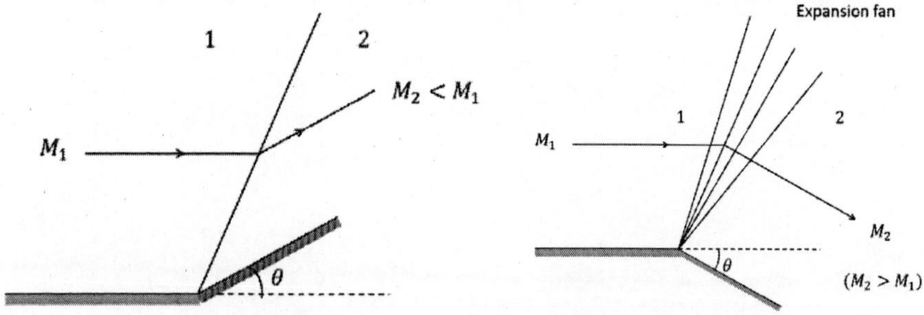

(a) Compression corner (b) Expansion corner

Figure 7.6: Supersonic Flow Over
(a) Compression Corner and (b) Expansion Corner

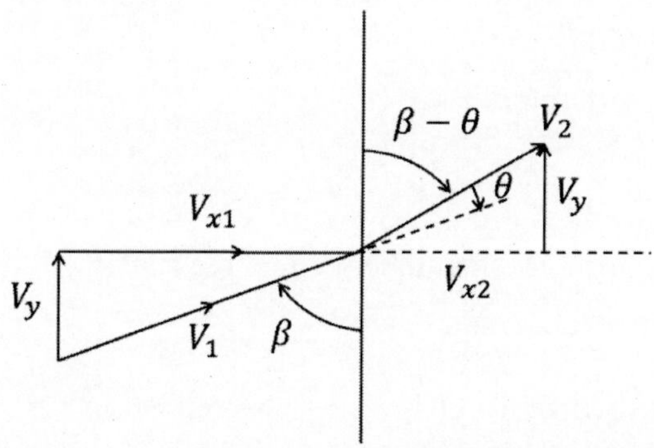

Figure 7.7: Flow Through an Oblique Shock Wave

Refer Figure 7.7. The velocity V_1 is having two components V_{x_1} and V_y. Therefore,

$$V_1 = \sqrt{V_{x_1}^2 + V_y^2}$$

and,

$$\tan\beta = \frac{V_{x_1}}{V_y}$$

The angle β is called shock wave angle.

Similarly, after the shock the velocity V_2 will have components V_{x_2} and V_y. The component V_{x_2} will always be less than V_{x_1} and the component V_y remains the same.

Therefore,

$$V_2 = \sqrt{V_{x_2}^2 + V_y^2}$$

and

$$\tan(\beta - \theta) = \frac{V_{x_2}}{V_y}$$

also,

$$M_1 = \frac{V_1}{a_1}$$

The component of M_1 normal to the shock,

$$M_{n_1} = M_1 \sin\beta$$

Apply normal shock relations on M_{n_1},

$$\frac{\rho_2}{\rho_1} = \frac{(\gamma+1)M_1^2 sin^2\beta}{(\gamma-1)M_1^2 sin^2\beta + 2}$$

$$\frac{p_2}{p_1} = 1 + \frac{2\gamma}{(\gamma+1)}(M_1^2 sin^2\beta - 1)$$

The normal component Mach number behind the shock will be,

$$M_{n_2}^2 = \frac{M_1^2 sin^2\beta + \frac{2}{(\gamma-1)}}{\frac{2\gamma}{(\gamma-1)} M_1^2 sin^2\beta - 1}$$

and the Mach number behind the oblique shock,

$$M_2 = \frac{M_{n_2}}{\sin(\beta - \theta)}$$

In the above equation,

$$M_2 = \frac{V_2}{a_2}$$

and,

$$M_{n_2} = \frac{V_{x_2}}{a_2}$$

The maximum value of β is that for a normal shock,

$$\beta = \frac{\pi}{2}$$

from continuity equation,

$$\frac{V_{x_2}}{V_{x_1}} = \frac{\rho_1}{\rho_2}$$

$$\frac{\tan(\beta - \theta)}{\tan\beta} = \frac{V_{x_2}}{V_{x_1}}$$

$$\frac{\tan(\beta - \theta)}{\tan\beta} = \frac{(\gamma - 1)M_1^2 sin^2\beta + 2}{(\gamma + 1)M_1^2 sin^2\beta}$$

This is called $\theta - \beta - M$ relation which is important for analyzing oblique shocks.

If M_1 is very large,

$$\beta = \frac{(\gamma + 1)}{2}.\theta$$

7.8 FANNO FLOW

Consider one dimensional steady flow through a constant area duct. Let the flow be adiabatic i.e., no external heat exchange and no external shaft work. Only the effect of viscosity i.e., friction is considered. Such a flow is designated as Fanno flow. An example of Fanno flow is aircraft propulsion engine. The energy equation is given as,

$$h + \frac{V^2}{2} = h_0$$

h_0 , the stagnation enthalpy and will have a constant value in all sections of the duct.

The continuity equation is given as,

$$\frac{\dot{m}}{A} = \rho V = G = \text{constant}$$

G is called mass velocity and will have constant value in all sections of the duct.

Eliminate V in the energy equation using continuity equation,

$$V = \frac{G}{\rho}$$

$$h + \frac{G^2}{2\rho^2} = h_0$$

$$h = h_0 - \frac{G^2}{2\rho^2}$$

As, h_0 and G are constant, this gives a relation between h and ρ.

It is known that entropy is a function of enthalpy and density. Therefore, an enthalpy-entropy diagram can be constructed. Figure 7.8 gives the line for a constant value of h_0 and G and this is called Fanno line.

From the Second law of thermodynamics, it is known that for an adiabatic flow, entropy should increase, and it cannot decrease and therefore the path of flow will be towards right. For a subsonic flow, due to friction (path 2 to a) the velocity and Mach number will increase, and enthalpy, temperature and pressure will decrease, and for a supersonic flow the velocity and Mach number will decrease, and enthalpy, temperature and pressure will increase. The limit of change is $M = 1$, point 'a' in the above diagram and therefore a supersonic flow may never become subsonic and vice-versa. The enthalpy at state 'a', where $M = 1$ can be calculated by differentiating the entropy equation with respect to h i.e., ds/dh and set it equal to zero. The stagnation enthalpy remains constant in Fanno flow. The flow in a constant area duct (Figure 7.9) along the length is analyzed to study the flow characteristics.

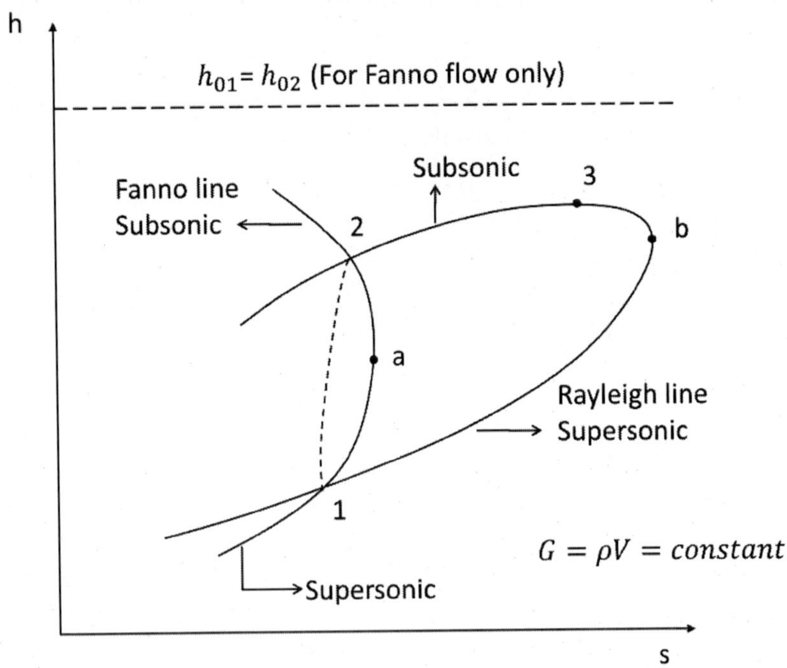

Figure 7.8: Fanno and Rayleigh Lines

(*Source:* Saravanamuttoo HIH, Rogers GFC and Cohen H., Gas Turbine Theory, Pearson Education, 2005)

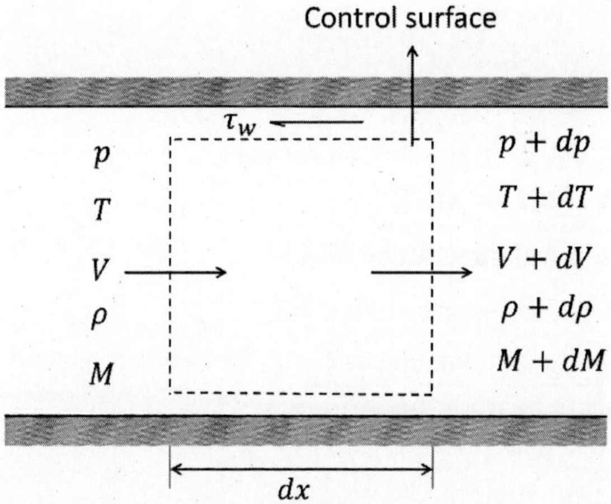

Figure 7.9: Control Surface for an Adiabatic Constant Area Fanno Flow

Consider the control volume as shown in the figure above and properties on upstream and downstream are given. τ_w is the shear stress which acts on the wall of the duct.

From equation of state, $p = \rho RT$, on differentiation,

$$\frac{dp}{p} = \frac{d\rho}{\rho} + \frac{dT}{T}$$

and Mach number,

$$M^2 = \frac{V^2}{\gamma RT}$$

take log and differentiate

$$\frac{dM^2}{M^2} = \frac{dV^2}{V^2} - \frac{dT}{T}$$

The energy equation

$$C_p dT + d\left(\frac{V^2}{2}\right) = 0$$

divide by C_pT and put the value of M

$$\frac{dT}{T} + \frac{\gamma - 1}{2} M^2 \frac{dV^2}{V^2} = 0$$

The continuity equation

$$\rho V = G = \text{constant}$$

take log and differentiate

$$\frac{d\rho}{\rho}+\frac{1}{2}\frac{dV^2}{V^2}=0$$

The momentum equation for the control volume can be written as,

$$-Adp-\tau_w\, dA_w=\dot{m}\, dV$$

dA_w is the wetted wall area of the duct.

The coefficient of friction, f is defined as,

$$f=\frac{\text{wall shearing stress}}{\text{dynamic head of the stream}}$$

$$=\frac{\tau_w}{\left(\frac{1}{2}\rho V^2\right)}$$

The hydraulic diameter D, is defined as,

$$D=\frac{4(\text{cross sectional area})}{\text{wetted perimeter}}$$

$$=\frac{4A}{dA_w/dx}$$

$$=4.\frac{A}{dA_w}.dx$$

The term hydraulic diameter is valid for ducts with non-circular cross sections.

Now, the momentum equation can be written as,

$$-dp-4f\frac{\rho V^2}{2}\frac{dx}{D}=\frac{\dot{m}}{A}dV=\rho V^2\frac{dV}{V}$$

divide by p and put,

$$\rho V^2=\gamma p M^2$$

$$\frac{dp}{p}+\frac{\gamma M^2}{2}.4f.\frac{dx}{D}+\frac{\gamma M^2}{2}.\frac{dV^2}{V^2}=0$$

Earlier, we have shown that for isentropic flow,

$$p_0=p\left(1+\frac{\gamma-1}{2}M^2\right)^{\gamma/\gamma-1}$$

take log and differentiate,

$$\frac{dp_0}{p}=\frac{dp}{p}+\frac{\gamma M^2/2}{1+\frac{\gamma-1}{2}M^2}.\frac{dM^2}{M^2}$$

A new parameter, impulse function F is defined as,

$$F = pA + \rho A V^2$$
$$= pA(1 + \gamma M^2)$$

take log and differentiate,

$$\frac{dF}{F} = \frac{dp}{p} + \frac{\gamma M^2}{(1 + \gamma M^2)} \cdot \frac{dM^2}{M^2}$$

Choosing the variable $\frac{4f dx}{D}$ as independent parameter, the equations can be solved simultaneously to calculate the other variables.

The friction has the net effect of accelerating a subsonic stream and causes a rise in static pressure at supersonic speeds.

For subsonic flow, density decreases, p_0 decreases and impulse function F decreases.

For supersonic flow, density increases, p_0 decreases and impulse function F decreases.

At point a on Fanno line (Figure 7.8), the flow is choked and further friction (i.e., additional length of pipe) will not serve any purpose and M will remain equal to unity.

7.9 RAYLEIGH FLOW

Consider a constant area duct and continuous change in flowing stream is due to energy effect. Combustion or moisture condensation are examples of energy effects. The stagnation enthalpy or stagnation temperature changes are involved in the flow. The frictional effects are neglected.

The momentum equation,

$$p + \rho V^2 = \frac{F}{A} = \text{constant}$$

and continuity equation,

$$\rho V = \frac{\dot{m}}{A} = G = \text{constant}$$

eliminate V,

$$p + \frac{G^2}{\rho} = \frac{F}{A}$$

This gives a relation between pressure and density as G and F are constant and therefore, a relationship between enthalpy and entropy can be obtained and curves can be drawn as shown in Figure 7.8. Such a line called Rayleigh line. In the curve of Rayleigh line, the maximum entropy is for subsonic flow and the curve below the maximum entropy point is for supersonic flow. The process of simple heating is thermodynamically reversible and therefore heat addition corresponds to entropy increase and heat rejection corresponds to entropy decrease. For subsonic flow, Mach number increases during heating and decreases during cooling and for supersonic flow, it is decreased during heating and increased during cooling. In all cases, the tendency of flow is to reach $M = 1$ (state b).

A typical flow behaviour between state 3 and b in subsonic flow is observed. At state 3, $M = 1 / \sqrt{\gamma}$ and when a subsonic flow is heated, the maximum temperature is reached. Between state 3 and b, the density decreases and hence the velocity increases, at such a rate that the static temperature falls. At point b, $M = 1$ and further heating will not serve any purpose and M will remain equal to unity.

When heat is added, for subsonic flow,

(a) Pressure decreases
(b) Mach number increases
(c) Velocity increases
(d) Temperature increases up to $M = 1 / \sqrt{\gamma}$ and then decreases
(e) Total temperature T_0 increases
(f) Total pressure decreases

For supersonic flow, when heat is added

(a) Pressure increases
(b) Mach number decreases
(c) Velocity decreases
(d) Temperature increases
(e) Total temperature T_0 increases
(f) Total pressure decreases

In practical situations variation in area, heat addition and friction may be present simultaneously.

SUMMARY

In this chapter, we have studied the basics of compressible fluid flow, perfect gas relationships, speed of sound wave and Mach number, isentropic flow, convergent-divergent (C-D) nozzle, shock waves, Fanno and Rayleigh flows and effect of frictional duct in subsonic and supersonic flow.

Some relationships which are useful in analyzing the practical problems are as follows:

For isentropic flow,

$$\frac{T_2}{T_1} = \left(\frac{p_2}{p_1}\right)^{(\gamma-1)/\gamma} = \left(\frac{\rho_2}{\rho_1}\right)^{(\gamma-1)}$$

Speed of sound wave,

$$a = \sqrt{\gamma RT}$$

Mach number,

$$M = \frac{V}{a}$$

Isentropic relations:

Ratios of stagnation (total) temperature, pressure and density are:

$$\frac{T_0}{T} = 1 + \frac{\gamma - 1}{2} M^2$$

$$\frac{p_0}{p} = \left(1 + \frac{\gamma - 1}{2} M^2\right)^{\gamma/(\gamma-1)}$$

$$\frac{\rho_0}{\rho} = \left(1 + \frac{\gamma - 1}{2} M^2\right)^{1/(\gamma-1)}$$

Properties at throat of the C-D nozzle, for air put $\gamma = 1.4$,

then,

$$T^* = 0.833\, T_0$$

$$p^* = 0.528\, p_0$$

$$\rho^* = 0.634\, \rho_0$$

for a normal shock,

$$V_1 V_2 = a^{*^2}$$

This is called the Prandtl relation.

In terms of speed ratio $M^* = V/a^*$

$$M_2^* = \frac{1}{M_1^*}$$

for normal shock,

$$T_{02} = T_{01}$$

Normal shocks are formed only in supersonic flow and flow is always subsonic after the shock.

Fanno flow and Rayleigh flow are discussed, and the governing equations are given. h-s diagrams are given for both friction and heat effect on fluid for subsonic and supersonic flow.

EXAMPLES

Example 7.1

For an adiabatic tank, air is entering with velocity of 150 m/s and coming out with 250 m/s and the temperature increase from inlet to exit is 20°C. Find the work done on air.

Solution

Data given:

$$V_1 = 150 \text{ m/s},$$
$$V_2 = 250 \text{ m/s},$$
$$T_2 - T_1 = 20 \text{ °C}$$

Calculate $W = ?$

Let $C_p = 1.005 \text{ kJ/kg.K}$

From energy equation,

$$h_1 + \frac{V_1^2}{2} + W = h_2 + \frac{V_2^2}{2} + Q$$

as, $h = C_p T$ and $Q = 0$,

$$C_p T_1 + \frac{V_1^2}{2} + W = C_p T_2 + \frac{V_2^2}{2}$$

$$W = C_p(T_2 - T_1) + \left(\frac{V_2^2 - V_1^2}{2}\right)$$

$$= 1.005 \times 20 \times 10^3 + \frac{250^2 - 150^2}{2}$$

$$= 20100 + 20000$$

$$= 40100\ J/kg$$

$$= \mathbf{40.1\ kJ/kg}$$

Example 7.2

A De Laval nozzle is to be designed for an exit Mach number of 1.5. The reservoir conditions are, $p_0 = 1$ atm (gauge), $T_0 = 20$ °C, $\gamma = 1.4$. Assuming shock free flow in the nozzle. Evaluate the exit absolute pressure in atm.

Solution

Data given:

$T_0 = 20\ °\text{C} + 273 = 293$ K, $\gamma = 1.4$, $p_0 = 1$ atm. (gauge) = 1 + 1 atm. = 2 atm absolute.

Isentropic relation,

$$\frac{p_0}{p} = \left(1 + \frac{\gamma - 1}{2} M^2\right)^{\gamma/(\gamma-1)}$$

$$p = \frac{p_0}{(1 + 0.2\,M^2)^{3.5}}$$

$$= \frac{p_0}{(1 + 0.2 \times (1.5)^2)^{3.5}}$$

$$= \frac{p_0}{(1 + 0.2 \times 2.25)^{3.5}}$$

$$= \frac{2}{(1.450)^{3.5}} = \frac{2}{3.671}$$

$$= \mathbf{0.5448\ atm.}$$

Example 7.3

An ideal conical diffuser has entry and exit diameter of 10 cm and 20 cm respectively. The pressure, temperature, and velocity of air at entry are 0.7 bar, 350 K and 200 m/s. Determine (a) exit pressure and exit velocity. Assume that the flow is incompressible.

Solution

Data given:

$p_1 = 0.7$ bar, $T_1 = 350$ K, $V_1 = 200$ m/s, $D_1 = 0.10$ m, $D_2 = 0.2$ m

Calculate: p_2, V_2

$$p_1 = \rho_1 R T_1$$

$$\rho_1 = \frac{p_1}{RT_1} = \frac{0.7 \times 1.0133 \times 10^5}{287 \times 350} = 0.7061\ \text{kg/m}^3$$

$$a_1 = \sqrt{\gamma R T_1} = \sqrt{1.4 \times 287 \times 350}$$
$$= 375 \text{ m/s}$$
$$M_1 = \frac{200}{375} = 0.53$$

therefore, density effect will not come.

therefore, $\rho_1 = \rho_2 =$ constant

continuity equation,

$$A_1 V_1 = A_2 V_2$$
$$V_2 = V_1 \frac{A_1}{A_2}$$
$$= 200 \times \frac{\frac{\pi}{4} \times (0.1)^2}{\frac{\pi}{4} \times (0.2)^2}$$
$$= \frac{200}{4}$$
$$= \mathbf{50 \ m/s}$$

From energy equation,

$$T_1 + \frac{V_1^2}{2C_p} = T_2 + \frac{V_2^2}{2C_p}$$
$$T_2 = T_1 + \frac{V_1^2 - V_2^2}{2C_p}$$
$$T_2 = 350 + \frac{200^2 - 50^2}{1.005 \times 10^3}$$
$$= 387.31 \ K$$
$$p_2 = 0.7061 \times 287 \times 387.31$$
$$= 78489.33$$
$$= \frac{78489.33}{1.0133 \times 10^5}$$
$$= \mathbf{0.77458 \ bar}$$

Example 7.4

The flow Mach number, pressure, and temperature ahead of a normal shock are given as 2.5, and 0.5 atm and 300 K respectively. Determine M_2, p_2, T_2 and V_2 behind the wave.

Solution

From the table given in Appendix- B (Normal shock in perfect gas, $\gamma = 1.4$) for $M_1 = 2.5$

$$\frac{p_2}{p_1} = 7.1250$$

$$\frac{T_2}{T_1} = 2.1375$$

$$\boldsymbol{M_2 = 0.5130}$$

therefore,

$$p_2 = 7.1250 \times 0.5 = \mathbf{3.5625\ atm}$$

$$T_2 = 2.1375 \times 300 = \mathbf{641.25\ K}$$

For $T_2 = 641.25\ K$

$$a_2 = \sqrt{1.4 \times 287 \times 641.25}$$
$$= 507.59\ m/s$$

thus,

$$V_2 = M_2.a_2$$
$$= 0.5130 \times 507.59$$
$$= \mathbf{260.39\ m/s}$$

Example 7.5

There is a normal shock in a uniform air stream. The properties of upstream of the shock are $V_1 = 450\,\text{m/s}$, $p_1 = 90\ kPa$ and $T_1 = 350\ K$. Determine p_2, V_2, T_2, T_{02} and p_{02} downstream of the shock.

Solution

$$T_1 = 350\ K$$

$$a_1 = \sqrt{\gamma R T_1}$$
$$= \sqrt{1.4 \times 287 \times 350}$$
$$= 375\ \text{m/s}$$

thus, $M_1 = V_1/a_1 = 450/375$

$$M_1 = 1.2$$

From table for normal shock in perfect gases, $\gamma = 1.4$ (see Appendix – B). For $M_1 = 1.2$

$$\frac{p_2}{p_1} = 1.5133$$

$$\frac{T_2}{T_1} = 1.1280$$

$$M_2 = 0.8422$$

thus,

$$p_2 = 1.5133 \times 90 = \mathbf{136.197\ kPa}$$

$$T_2 = 1.1280 \times 350 = \mathbf{394.8\ K}$$

thus,

$$a_2 = \sqrt{\gamma R T_2}$$

$$= \sqrt{1.4 \times 287 \times 394.8}$$

$$= 398.28\ \mathrm{m/s}$$

therefore,

$$V_2 = M_2 \times a_2$$

$$= 0.8422 \times 398.28$$

$$= \mathbf{335.43\ m/s}$$

There is no heat added to or taken away from the flow as it transfers the shock wave i.e., the flow process across the shock wave is adiabatic. Therefore,

$$T_{02} = T_{01}$$

$$\frac{T_{01}}{T_1} = \left(1 + \frac{\gamma - 1}{2} M_1^2\right)$$

$$T_{01} = 350\left(1 + \frac{1.4 - 1}{2} . (1.20)^2\right)$$

$$= 350(1 + 0.2 \times (1.20)^2)$$

$$= 350 \times (0.288 + 1)$$

$$= 450.8\ \mathrm{K}$$

therefore, $\boldsymbol{T_{02}} = \mathbf{450.8\ K}$

$$\frac{p_{01}}{p_1} = \left(1 + \frac{\gamma - 1}{2} M_1^2\right)^{(\gamma/(\gamma-1))} = (1 + 0.2 \times (1.20)^2)^{(1.4/0.4)}$$

$$= 2.42$$

$$p_{01} = 90 \times 2.42 = 217.8\ Pa$$

From table given in Appendix-B, for $M_1 = 1.2$

$$\frac{p_{02}}{p_{01}} = 0.9928$$

therefore,

$$p_{02} = 0.9928 \times 217.8\ kPa$$
$$= \mathbf{216.23\ kPa}$$

QUESTIONS

1. Derive the expression for speed of sound wave and define Mach number.
2. Show that the area-velocity relation for one dimensional compressible flow is given as:
$$\frac{dA}{dV} = -\frac{A}{V}(1 - M^2)$$
3. Discuss the pressure profile along the length of the C-D nozzle under varying values of back pressure.
4. Define Fanno flow and draw its T-s diagram. Explain the effect of friction on various properties for subsonic and supersonic flows.
5. Give the governing equations for Rayleigh flow and draw its T-s diagram. Explain the effect of heating on fluid properties for subsonic and supersonic flows.

CHAPTER 8

Intake and Propelling Nozzle

8.1 INTRODUCTION

The purpose of inlet is to reduce the velocity and increase the pressure of the free stream air intake to the conditions required at the entrance of fan and/or compressor with minimum of total pressure loss and drag force as small as possible. The performance of compression is best with a uniform flow of air at a Mach number of about 0.5. The design of the intake influences engine efficiency and aircraft safety. Non-uniform or distorted flow may cause compressor surge and may result in flame blow out or severe mechanical damage. Some flow distortion is unavoidable during rapid maneuvering of subsonic aircraft at Mach 0.8 to 0.85, while supersonic aircraft may operate at Mach 2 to 2.5. At take off with zero forward speed, the engine will operate at maximum power and airflow. The intake works in a wide range of operating conditions.

The purpose of the propelling/exhaust nozzle is to increase the velocity of the exhaust gas before discharge from the nozzle and to collect and straighten the gas flow. For large thrust, the kinetic energy of the exhaust gas must be high and requires a high exhaust velocity. The pressure across the nozzle controls the expansion process. The maximum thrust is obtained when the exit pressure equals the ambient pressure. The two basic nozzles used in jet engines are convergent and convergent-divergent (C-D) nozzles.

8.2 INTAKE

The purpose of intake duct is to supply sufficient air to the compressor with minimum losses. In most subsonic aircraft the engine is placed in nacelles. Figure 8.1 gives the streamline patterns of a subsonic inlet.

The operating conditions of inlet depend on the flight velocity and the maximum flow requirements. The flow within the inlet undergoes diffusion in a divergent duct. This reduction in flow velocity creates an increase in static pressure.

Figure 8.2 gives the schematic diagram of a typical supersonic inlet. Supersonic inlets are classified by their percentage of internal compression. Internal compression refers to the amount of supersonic area change that takes place between the cowl lip and the throat. The total supersonic area change is the difference between the capture area and the throat area. The area change that occurs in front of the cowl lip is called the external compression. Inlets are usually classified as external compression inlet, mixed compression inlet and internal compression inlets.

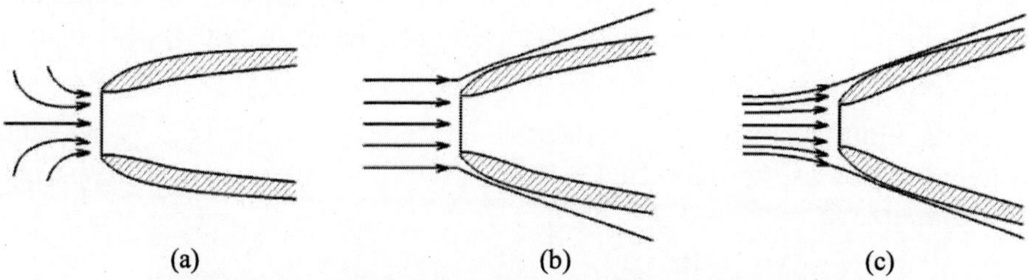

Figure 8.1: Typical Streamline Patterns for Subsonic Inlet, (a) Static Operation (insufficient air) (b) Low Speed Operation (correct air) (c) High Speed Operation (more air than the engine needs)

(*Source:* Mattingly, Jack D., Elements of Gas Turbine Propulsion, McGraw Hill Education, 2017)

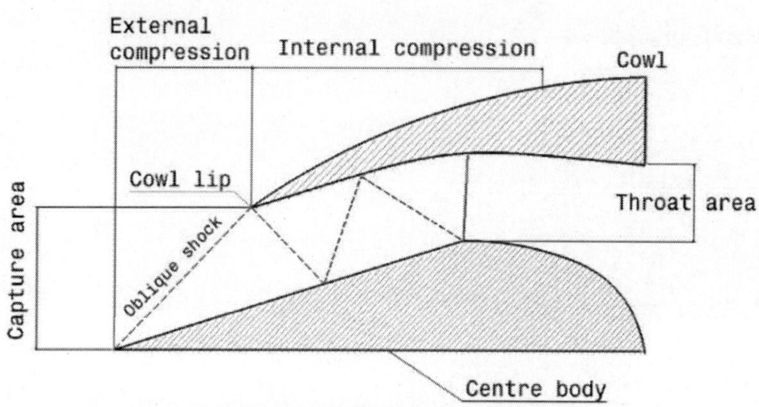

Figure 8.2: Supersonic Inlet

(*Source:* Ganesan V., Gas Turbines, Tata McGraw Hill Publishing Company Limited, Second Edition, 2007)

The characteristics of a supersonic inlet are:

1. Provide adequate subsonic performance,
2. good pressure distribution at the compression inlet,
3. high pressure recovery rates, and
4. must be able to operate efficiently at all ambient pressures and temperatures during take-off, subsonic flight, as well as its supersonic design condition.

Consider intake as an adiabatic duct. With no heat or work transfer, the stagnation temperature will remain constant. There will be loss of stagnation pressure due to friction and due to shock waves at supersonic flight speeds. Under static conditions or at very low speeds, the intake will work as a nozzle and at normal forward speed it will work as a diffuser changing speed from V_a to V_1 and the static pressure will rise from p_a to p_1. Suffix a is for ambient conditions and 1 is for exit of the nozzle. The pressure rise $(p_{01} - p_a)$ is important and is called ram pressure rise.

The intake efficiency may be defined as an isentropic efficiency η_i, in terms of temperature rise or ram efficiency η_r, in terms of pressure rise. Applying energy equation,

$$h_{01} = h_{0a} = h_a + \frac{V_a^2}{2}$$

h_0 is total enthalpy, as $h = C_p T$

$$T_{01} = T_{0a} = T_a + \frac{V_a^2}{2C_p}$$

From isentropic relation,

$$\frac{p_{01}}{p_a} = \left(\frac{T'_{01}}{T_a}\right)^{\gamma/(\gamma-1)}$$

Figure 8.3 shows the T-s diagram and sketch of intake nozzle.

The isentropic efficiency, η_i is defined as,

$$\eta_i = \frac{(T'_{01} - T_a)}{(T_{01} - T_a)}$$

$$(T'_{01} - T_a) = (T_{01} - T_a)\eta_i$$

$$= \eta_i . \frac{V_a^2}{2C_p}$$

$$\frac{T'_{01}}{T_a} - 1 = \eta_i . \frac{V_a^2}{2C_p . T_a}$$

$$\frac{T'_{01}}{T_a} = 1 + \eta_i . \frac{V_a^2}{2C_p . T_a}$$

therefore,

$$\frac{p_{01}}{p_a} = \left(1 + \eta_i . \frac{V_a^2}{2C_p . T_a}\right)^{\gamma/(\gamma-1)}$$

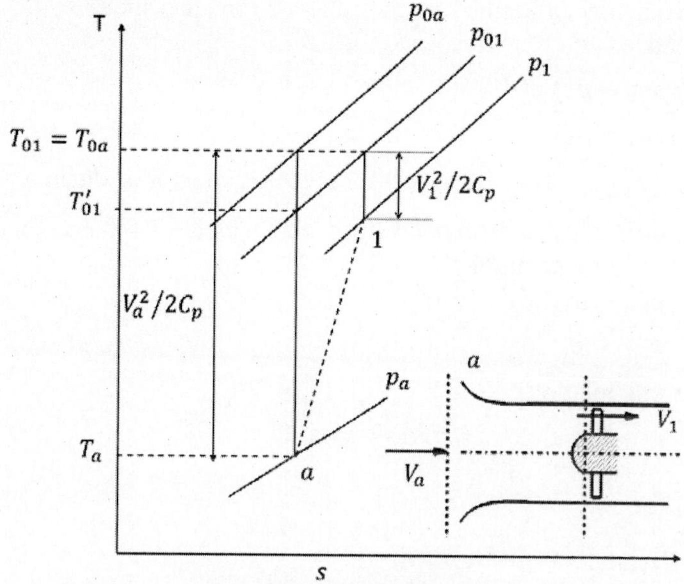

Figure 8.3: The Intake Loss
(*Source:* Saravanamuttoo HIH, Rogers GFC and Cohen H, Gas Turbine Theory, 5th Edition, 2005)

also, velocity of sound, $a = \sqrt{\gamma RT}$ and Mach number, $M = V/a$

therefore,

$$M_a = \frac{V_a}{\sqrt{\gamma RT_a}}$$

also,

$$C_p = \frac{\gamma R}{(\gamma - 1)}$$

we get,

$$\frac{p_{01}}{p_a} = \left(1 + \eta_i . \frac{\gamma RT_a M_a^2 . (\gamma - 1)}{2 . T_a . \gamma R}\right)^{\gamma/(\gamma-1)}$$

$$= \left(1 + \eta_i . \frac{(\gamma - 1)}{2} M_a^2\right)^{\gamma/(\gamma-1)}$$

and,

$$\frac{T_{01}}{T_a} = \left(1 + \frac{V_a^2}{2C_p T_a}\right) = \left(1 + \frac{(\gamma-1)}{2} M_a^2\right)$$

The ram efficiency is defined as the ratio of ram pressure rise to the inlet dynamic head.

$$\eta_r = \frac{(p_{01} - p_a)}{(p_{0a} - p_a)}$$

The term, p_{01}/p_{0a} is also called pressure recovery factor of the intake.

The intake efficiency will depend upon the location of the engine in the aircraft, viz. in wing or fuselage.

$$\frac{p_{01}}{p_a} = \frac{p_{01}}{p_{0a}} \times \frac{p_{0a}}{p_a}$$

From isentropic relation,

$$\frac{p_{0a}}{p_a} = \left[1 + \left(\frac{\gamma - 1}{2}\right) M_a^2\right]^{\gamma/(\gamma-1)}$$

8.3 PROPELLING NOZZLE

In propelling or exhaust nozzle, the hot products of combustion are expanded to give a high velocity jet. In a simple jet engine, there will be a single nozzle downstream of the turbine. The turbofan may have two separate nozzles for hot and cold streams or, the hot and cold streams may be mixed and leave from a single nozzle. There will be friction losses in the exit jet pipe. An afterburner is incorporated, when thrust boosting is required, and is installed in the jet pipe.

The functions of an exhaust nozzle are as given below:

1. Accelerate the flow to a high velocity with minimum total pressure loss.
2. Match exit and atmospheric pressures as closely as desired.
3. Permit afterburner operation without affecting main engine operation. This function requires a variable area nozzle.
4. Allow for cooling of walls.
5. Mix core and bypass turbofan, if required.
6. Allow for thrust reversing, if desired.
7. Suppress jet noise, if desired.
8. Thrust vector control, if desired.

The convergent nozzle is a simple convergent duct as shown in Figure 8.4 below. This is used for low total nozzle pressure ratio, less than about 4. This nozzle is generally used for subsonic aircraft.

The convergent-divergent (C-D) nozzle has two sections (i) convergent duct and (ii) divergent duct. The section where the cross-sectional area is minimum is called the throat. Most C-D nozzles incorporate variable geometry and other aerodynamic features. These are used when the total nozzle pressure ratio is high, greater than about 6. High performance engines in supersonic aircraft employ C-D nozzles. The nozzle also serves as a back pressure control for the engine and an acceleration device converting gas thermal energy to kinetic energy. A secondary function of the nozzle is to provide required thrust reversing or thrust vectoring. The throat area of the nozzle is one of the means to control the thrust and fuel consumption. Changing the nozzle throat area from its original design value and the operating characteristics will change the performance both on and off design conditions.

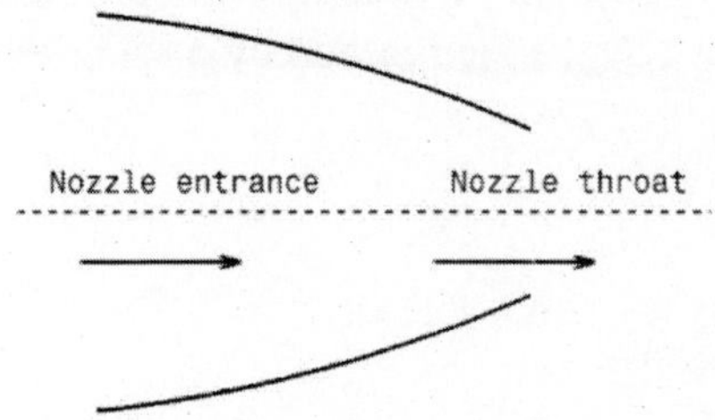

Figure 8.4: Convergent Propelling Nozzle

Large changes in exhaust nozzle throat area are required for afterburning engines to compensate for the large changes in total temperature leaving the afterburner. Opening the nozzle throat area to its maximum value reduces the back pressure on the turbine and increases its expansion ratio.

Thrust reversers are used in aircraft to supplement the brakes. Two basic types of thrust reversers are used: the cascade blocker type and the clamshell type, shown in Figure 8.5.

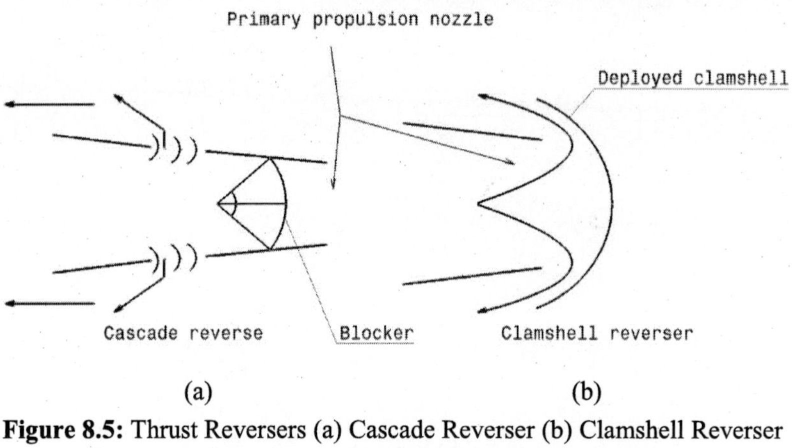

Figure 8.5: Thrust Reversers (a) Cascade Reverser (b) Clamshell Reverser

(*Source:* Mattingly, Jack D., Elements of Gas Turbine Propulsion, McGraw Hill Education, 2017)

In the cascade blocker type, the primary nozzle is blocked off, and cascades are opened in the upstream portion of the nozzle duct to reverse the flow. In the clamshell type, the exhaust gas is split and reversed by the clamshell.

Thrust vectoring is required in combat aircraft. Vectoring nozzles have been used in vertical take-off and landing (VTOL) aircraft which helps in improving maneuvering and augmenting lift.

Figure 8.6 gives the T-s diagram for a convergent nozzle. p_{04} and T_{04} are the total pressure and total temperature at entry to the nozzle and p_5 and T_5 are the static pressure and temperature at exit of the nozzle. The critical pressure p_c is estimated from the isentropic relation by putting $M = 1$.

$$\frac{p_0}{p} = \left(1 + \frac{\gamma - 1}{2} M^2\right)^{\gamma/(\gamma-1)}$$

For $M = 1, p = p_c$,

$$\frac{p_0}{p_c} = \left(\frac{\gamma + 1}{2}\right)^{\gamma/(\gamma-1)}$$

Put $\gamma = 1.33$ for hot gases,

$$\frac{p_0}{p_c} = 1.853$$

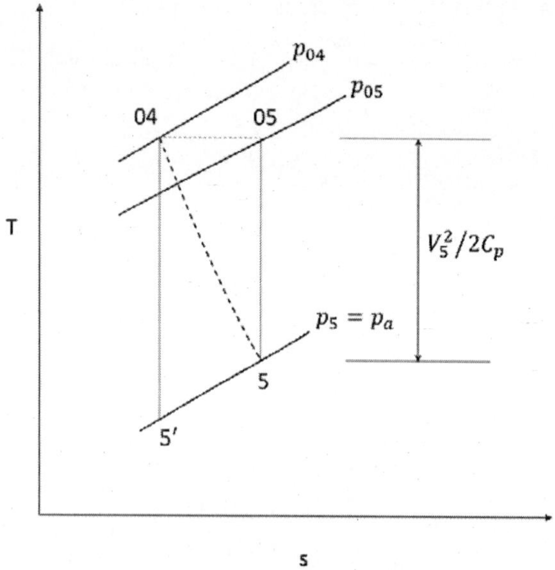

(a) When $p_{04}/p_a < p_{04}/p_c$,
p_a = atmospheric pressure, T_c = Temperature for choked flow

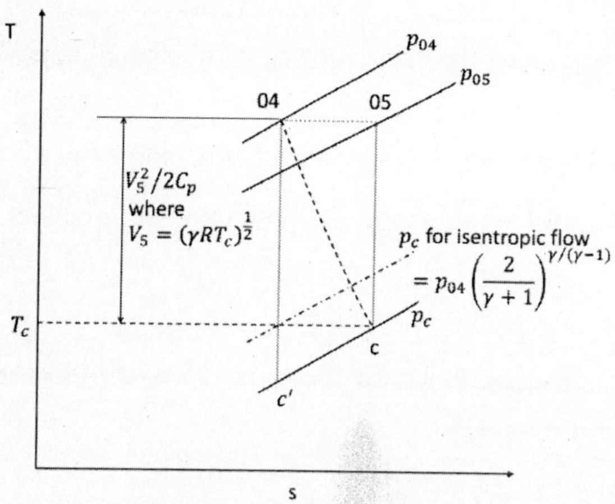

(b) When $p_{04}/p_a > p_{04}/p_c$,

Figure 8.6: Nozzle Loss for (a) Unchoked and (b) Choked Flow

Two cases will arise, one when the expansion to p_a is complete in the nozzle, i.e., when $p_{04}/p_a < p_{04}/p_c$ and the flow is unchoked and the other is when $p_{04}/p_a > p_{04}/p_c$ and the flow is choked.

The jet efficiency, η_j is defined as,

$$\eta_j = \frac{T_{04} - T_5}{T_{04} - T_5'}$$

$$T_{04} - T_5 = \eta_j (T_{04} - T_5')$$

$$= \eta_j \, T_{04}(1 - \frac{T_5'}{T_{04}})$$

From isentropic relation,

$$\frac{T_{05}}{T_5'} = \frac{T_{04}}{T_5'} = \left(\frac{p_{04}}{p_5}\right)^{(\gamma-1)/\gamma}$$

Therefore,

$$T_{04} - T_5 = \eta_j . T_{04}\left[1 - \frac{1}{\left(\frac{p_{04}}{p_5}\right)^{(\gamma-1)/\gamma}}\right]$$

$$= \frac{V_5^2}{2C_p}$$

$$(as, T_{05} = T_{04})$$

where, V_5 = velocity of the jet

For pressure ratios up to critical value, p_5 will be equal to p_a and the pressure thrust will be zero.

Above the critical pressure the nozzle is choked and $V_5 = a_5 = \sqrt{\gamma R T_5}$.

$T_{04} = T_{05}$ is valid for all adiabatic flow with no work transfer.

therefore,

$$\frac{T_{04}}{T_5} = \frac{T_{05}}{T_5} = 1 + \frac{V_5^2}{2C_pT_5}$$

$$= 1 + \frac{\gamma - 1}{2} M_5^2$$

Now, put $M_5 = 1$, $T_5 = T_c$

$$\frac{T_{04}}{T_c} = \frac{\gamma + 1}{2}$$

Again, refer Figure 8.6 (b), $T_{04} = T_{05}$

$$\eta_j = \frac{T_{04} - T_c}{T_{04} - T_c'}$$

$$T_{04} - T_c' = \frac{1}{\eta_j}(T_{04} - T_c)$$

$$T_c' = T_{04} - \frac{1}{\eta_j}(T_{04} - T_c)$$

From isentropic relation,

$$\frac{p_c}{p_{04}} = \left(\frac{T_c'}{T_{04}}\right)^{\gamma/(\gamma-1)}$$

$$p_c = p_{04}\left(\frac{T_c'}{T_{04}}\right)^{\gamma/(\gamma-1)}$$

$$= p_{04}\left[\frac{1}{T_{04}}\left\{T_{04} - \frac{1}{\eta_j}(T_{04} - T_c)\right\}\right]^{\gamma/(\gamma-1)}$$

$$= p_{04}\left[1 - \frac{1}{\eta_j}\left(1 - \frac{T_c}{T_{04}}\right)\right]^{\gamma/(\gamma-1)}$$

$$= p_{04}\left[1 - \frac{1}{\eta_j}\left(1 - \frac{2}{(\gamma + 1)}\right)\right]^{\gamma/(\gamma-1)}$$

$$= p_{04}\left[1 - \frac{1}{\eta_j}\left(\frac{\gamma - 1}{\gamma + 1}\right)\right]^{\gamma/(\gamma-1)}$$

$$\frac{p_{04}}{p_c} = \frac{1}{\left[1 - \frac{1}{\eta_j}\left(\frac{\gamma-1}{\gamma+1}\right)\right]^{\gamma/(\gamma-1)}}$$

Pressure thrust $= A_5(p_c - p_a)$

where, A_5 = Area at the exit of the nozzle and can be estimated from the mass flow rate, $\dot{m}_5$

$$A_5 = \frac{\dot{m}_5}{\rho_c . V_c}$$

and density, ρ_c can be estimated from

$$p_c = \rho_c R T_c$$

SUMMARY

Intake reduces the inlet air velocity and increases the pressure of the freestream air to the conditions required at the entrance of fan and/or compressor with minimum pressure loss. At take-off with zero forward speed, the engine operates at maximum power and airflow. The intake works in a wide range of operating conditions. The intake efficiency is defined as an isentropic efficiency in terms of total temperature rise or ram efficiency in terms of total pressure rise.

The purpose of propelling/exhaust nozzle is to increase the velocity of the exit gas before discharge from the nozzle and to collect and straighten the gas flow. The maximum thrust is obtained when the exit gas pressure equals atmospheric pressure. The two basic nozzles used in jet engines are convergent nozzle and convergent- divergent (CD) nozzle.

QUESTIONS

1. Briefly explain the characteristics of subsonic inlets.
2. What are the characteristics of supersonic inlets? Name the parameters by which assessment of performance can be done.
3. Define propelling nozzle. Which nozzle is used in supersonic aircraft jet engines? Justify the answer.
4. Explain the use of choked and unchoked propelling nozzle with the help of T-s diagram.

CHAPTER 9

Ideal Jet Engine Cycle Analysis

9.1 INTRODUCTION

The working fluid undergoes thermodynamic changes as it flows through the engine. There are two types of cycle analysis, (i) parametric cycle analysis, also called design point or on-design, and (ii) engine performance analysis, also called off-design. In parametric cycle analysis, performance of engines at different flight conditions are studied. In engine performance analysis, the performance of a specific engine at all flight conditions and throttle settings are studied. In parametric cycle analysis, the performance parameters, thrust and thrust specific fuel consumption are determined which helps in selecting the type of engine for a particular need. The design choices are compressor pressure ratio, fan pressure ratio, by-pass ratio, etc.; the design limitations are burner exit temperature, compressor exit pressure, etc. and the flight environment, i.e., the Mach number, ambient temperature, etc.

The total or stagnation temperature will be denoted as T_0, which is defined as the temperature when the fluid is brought to rest isentropically. The velocity is denoted as V. The symbol a is for velocity of sound.

For a calorically perfect gas,

$$T_0 = T + \frac{V^2}{2C_p}$$

where, T is static temperature.

A calorically perfect gas is a perfect gas with constant specific heat, C_p and C_v.

Mach number M,

$$M = \frac{V}{a}$$

$$a = \sqrt{\gamma RT}, C_p = \gamma R/(\gamma - 1), \gamma = C_p/C_v$$

therefore,

$$T_0 = T\left(1 + \frac{\gamma - 1}{2}M^2\right)$$

The total or stagnation pressure is defined as the pressure reached when a steady flow is brought to rest adiabatically and reversibly (i.e., isentropic).

From isentropic relations:

$$\frac{p_0}{p} = \left(\frac{T_0}{T}\right)^{\gamma/(\gamma-1)}$$

$$p_0 = p\left(1 + \frac{\gamma - 1}{2} M^2\right)^{\gamma/(\gamma-1)}$$

where, p_0 is total pressure and p is static pressure.

The total pressure ratio at exit and entry is denoted as π and total temperature ratio is represented as τ. For any component, the ratios are as follows:

$$\pi = \frac{\text{total pressure leaving the component}}{\text{total pressure entering the component}}$$

and,

$$\tau = \frac{\text{total temperature leaving the component}}{\text{total temperature entering the component}}$$

The subscripts for various components are, *d* for inlet (diffuser), *c* for compressor, *b* for burner/combustor, *t* for turbine, *n* for nozzle and *F* for fan.

For free stream, total/static temperature, and pressure ratios, τ_r and π_r respectively are defined as,

$$\tau_r = \frac{T_{0a}}{T_a} = \left(1 + \frac{\gamma - 1}{2} M_a^2\right)$$

$$\pi_r = \frac{p_{0a}}{p_a} = \left(1 + \frac{\gamma - 1}{2} M_a^2\right)^{\gamma/(\gamma-1)}$$

Suffix r, stands for free stream and a for ambient conditions.

T_{0a} and T_a are total and static temperature of the free stream i.e., ambient conditions. Similarly, p_{0a} and p_a are total and static pressure of the free stream. M_a is free stream Mach number.

The ratio of burner exit enthalpy, $C_p T_{04}$ to the ambient enthalpy, $C_p T_a$ is defined as, τ_λ.

$$\tau_\lambda = \frac{\left(C_p T_{04}\right)_{burner\ exit}}{C_p T_a}$$

The schematic diagram of gas turbine engines is shown in Figure 9.1 below.

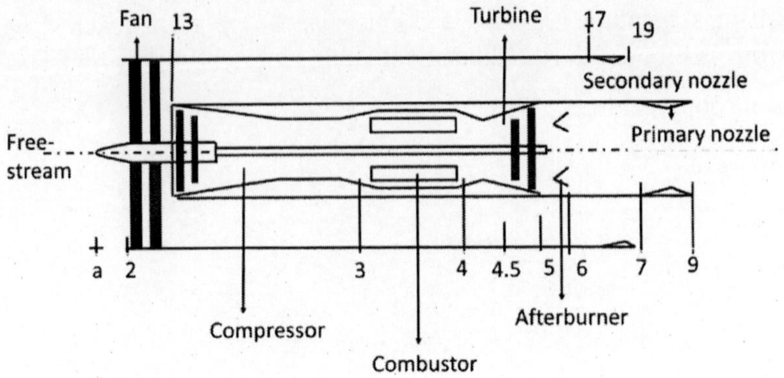

Figure 9.1: Station Numbering for Gas Turbine Engines
(*Source:* Mattingly, Jack D., Elements of Gas Turbine Propulsion, McGraw Hill Education, 2017)

For a jet engine with a single inlet and single exhaust, the following steps are involved in engine parametric cycle analysis. Engine performance will be determined by specific thrust and thrust specific fuel consumption for various flight conditions, such as Mach number under given design limits and component performance efficiencies. Station number a-9 for a core engine flow will be used. Station a is free stream and station 9 is, end of the propelling nozzle. The various components are:

- $a-2$ diffuser (inlet) (d)
- $2-3$ compressor (c)
- $3-4$ combustor/burner (b)
- $4-5$ turbine (t)
- $5-7$ afterburner (AB)
- $7-9$ nozzle (n)

Stream 13 to 19 is cold air/fan stream

$$M_a = \frac{V_a}{a_a},\ M_9 = \frac{V_9}{a_9}$$

$$a_a = \sqrt{\gamma R T_a},\ \ a_9 = \sqrt{\gamma R T_9}$$

A_9 = Area at exit of the nozzle
p_a = ambient pressure of free stream
p_9 = pressure at exit of the nozzle
T_a = Temperature of the free stream
T_9 = Temperature at exit of the nozzle

Similarly, temperature and pressure at various stations will be denoted.

Steps:

1. Uninstalled engine thrust, F

$$F = (\dot{m}_9 V_9 - \dot{m}_a V_a) + A_9(p_9 - p_a)$$

$\dot{m}_a$ = inlet air mass flow rate

$\dot{m}_9$ = mass flow rate in nozzle

$\dot{m}_f$ = mass flow rate of fuel injected in burner/combustor

$$\dot{m}_9 = \dot{m}_a + \dot{m}_f$$

$$\frac{F}{\dot{m}_a} = a_a\left(\frac{\dot{m}_9 V_9}{\dot{m}_a a_a} - M_a\right) + \frac{A_9 p_9}{\dot{m}_a}\left(1 - \frac{p_a}{p_9}\right)$$

2. Velocity ratio (V_9/a_a) in terms of Mach number, M_9.

$$\left(\frac{V_9}{a_a}\right)^2 = \frac{a_9^2 M_9^2}{a_a^2}$$

$$= \frac{\gamma_9 . R_9 . T_9}{\gamma_a . R_a . T_a} . M_9^2$$

3. Exit Mach number, M_9

$$p_{09} = p_9\left(1 + \frac{\gamma - 1}{2} M_9^2\right)^{\gamma/(\gamma-1)}$$

therefore,

$$M_9^2 = \frac{2}{(\gamma - 1)}\left[\left(\frac{p_{09}}{p_9}\right)^{(\gamma-1)/\gamma} - 1\right]$$

write, p_{09}/p_9 as

$$\frac{p_{09}}{p_9} = \frac{p_a}{p_9} . \frac{p_{0a}}{p_a} . \frac{p_{02}}{p_{0a}} . \frac{p_{03}}{p_{02}} . \frac{p_{04}}{p_{03}} . \frac{p_{05}}{p_{04}} . \frac{p_{07}}{p_{05}} . \frac{p_{09}}{p_{07}}$$

$$= \frac{p_a}{p_9} . \pi_r . \pi_d . \pi_c . \pi_b . \pi_t . \pi_{AB} . \pi_n$$

4. Write temperature ratio as

$$\frac{T_9}{T_a} = \frac{T_{09}/T_a}{T_{09}/T_9}$$

$$= \frac{T_{09}/T_a}{(p_{09}/p_9)^{(\gamma-1)/\gamma}}$$

$$\frac{T_{09}}{T_a} = \frac{T_{0a}}{T_a} . \frac{T_{02}}{T_{0a}} . \frac{T_{03}}{T_{02}} . \frac{T_{04}}{T_{03}} . \frac{T_{05}}{T_{04}} . \frac{T_{07}}{T_{05}} . \frac{T_{09}}{T_{07}}$$

$$= \tau_r . \tau_d . \tau_c . \tau_b . \tau_t . \tau_{AB} . \tau_n$$

5. Enthalpy balance in combustor/burner

   $\dot{m}_a C_p T_{03} + \dot{m}_f . h_{PR} = (\dot{m}_a + \dot{m}_f) C_p T_{04}$

   where, h_{PR} = lower heating value of fuel per unit mass.
6. Equate the turbine power output to compressor power and power required for operation of fan, propeller, etc.
7. Calculate specific thrust, thrust specific fuel consumption and thermal and propulsive efficiencies.

Assumptions of ideal cycle:

1. All compression, expansion, inlet (diffuser), fan, nozzle processes are isentropic i.e., reversible, and adiabatic. Therefore,

   $\tau_d = \tau_n = 1$

   $\pi_d = \pi_n = 1$

   $\tau_c = (\pi_c)^{(\gamma-1)/\gamma}$

   $\tau_t = (\pi_t)^{(\gamma-1)/\gamma}$
2. Constant pressure burning of fuel in the combustion chamber ($\pi_b = 1$). Fuel flow rate is much smaller with respect to air mass flow rate and hence neglected.

   $\dot{m}_f \ll \dot{m}_a$

   $\dot{m}_9 = \dot{m}_a + \dot{m}_f$

   $\simeq \dot{m}_a$
3. The working fluid is air and will be treated as perfect gas with constant specific heats.
4. The engine exhaust nozzles expand the gas to the ambient pressure ($p_9 = p_a$).

9.2 IDEAL RAMJET

The schematic diagram of a typical ramjet engine is given in Figure 9.2 and Figure 9.3 gives the ideal cycle on T-s diagram.

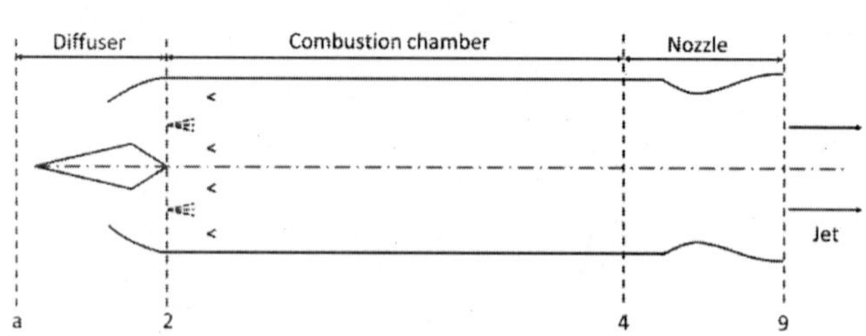

Figure 9.2: Sketch of a Ramjet Engine

The station details are:

- $a-2$ Diffuser (inlet)
- $2-4$ Combustion chamber
- $4-9$ Propelling nozzle

The pressure at diffuser outlet and combustion chamber is the same. The jet outlet pressure is the same as atmospheric pressure.

$$p_{02} = p_{04}, p_a = p_9$$

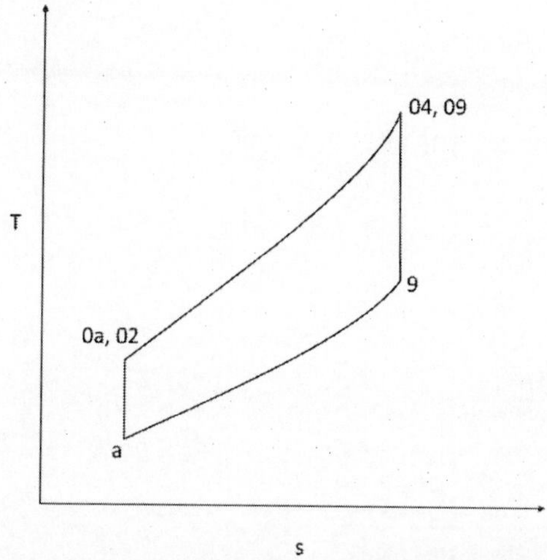

Figure 9.3: The T-s Diagram for an Ideal Ramjet Engine
(*Source:* Mattingly, Jack D., Elements of Gas Turbine Propulsion, McGraw Hill Education, 2017)

A ramjet is the simplest aircraft engine consisting of an inlet (diffuser), combustion chamber (burner) and a propelling nozzle. In inlet, the flight velocity V_a is reduced to a value V_2 resulting in increase of static pressure to p_2 and static temperature to T_2. In combustion chamber the fuel is injected and chemical energy of the fuel is converted to thermal energy increasing the temperature of products of combustion to T_4. Combustion takes place at near constant pressure. The products of combustion are expanded in nozzle. There is no compressor and hence no turbine in a ramjet engine. The cycle analysis is as follows:

Steps:

1. Uninstalled engine thrust

   $$F = (\dot{m}_9 V_9 - \dot{m}_a V_a) + A_9(p_9 - p_a)$$

   For an ideal engine,

   $$p_9 = p_a, \dot{m}_9 \cong \dot{m}_a$$

thus,

$$F = \dot{m}_a(V_9 - V_a)$$
$$= \dot{m}_a a_a\left(\frac{V_9}{a_a} - M_a\right)$$

2.

$$\left(\frac{V_9}{a_a}\right)^2 = \frac{a_9^2 M_9^2}{a_a^2}$$
$$= \frac{\gamma_9 R_9 T_9 M_9^2}{\gamma_a R_a T_a}$$

For an ideal engine, it is assumed that,

$$\gamma_9 = \gamma_a = \gamma$$
$$R_9 = R_a = R$$

therefore,

$$\left(\frac{V_9}{a_a}\right)^2 = \frac{T_9}{T_a}.M_9^2$$

3.

$$p_{09} = p_a.\frac{p_{0a}}{p_a}.\frac{p_{02}}{p_{0a}}.\frac{p_{04}}{p_{02}}.\frac{p_{09}}{p_{04}}$$
$$= p_a.\pi_r.1.\pi_b.1$$

as, $p_{02} = p_{0a}$ and $p_{04} = p_{09}$

For an ideal engine,

$$\pi_b = 1$$

therefore,

$$p_{09} = p_a.\pi_r$$
$$\frac{p_{09}}{p_9} = \frac{p_{09}}{p_a}.\frac{p_a}{p_9}$$
$$(as, p_{09} = p_{0a})$$
$$\frac{p_{09}}{p_9} = \pi_r.\frac{p_a}{p_9}$$

as, $p_9 = p_a$

$$\frac{p_{09}}{p_9} = \pi_r$$

$$M_9^2 = \frac{2}{(\gamma-1)}\left[\left(\frac{p_{09}}{p_9}\right)^{(\gamma-1)/\gamma} - 1\right]$$

$$= \frac{2}{(\gamma-1)}\left[(\pi_r)^{(\gamma-1)/\gamma} - 1\right]$$

$as, (\pi_r)^{(\gamma-1)/\gamma} = \tau_r$

therefore,

$$M_9^2 = \frac{2}{(\gamma-1)}(\tau_r - 1) = M_a^2$$

therefore, $M_9 = M_a$

Thus, for an ideal ramjet engine, the jet exit Mach number is same as the inlet air free stream Mach number.

4.

$$T_{09} = T_a.\frac{T_{0a}}{T_a}.\frac{T_{02}}{T_{0a}}.\frac{T_{04}}{T_{02}}.\frac{T_{09}}{T_{04}}$$

$$= T_a.\tau_r.1.\tau_b.1$$

$$= T_a.\tau_r.\tau_b.$$

also,

$$\frac{T_{09}}{T_9} = \left(\frac{p_{09}}{p_9}\right)^{(\gamma-1)/\gamma}$$

$$\frac{T_9}{T_a} = \frac{T_{09}/T_a}{T_{09}/T_9} = \frac{(T_{09}/T_{0a}).\frac{T_{0a}}{T_a}}{T_{09}/T_9}$$

$$= \frac{(T_{04}/T_{02}).\frac{T_{0a}}{T_a}}{T_{09}/T_9}$$

$$= \frac{\tau_b.\tau_r}{(p_{09}/p_9)^{(\gamma-1)/\gamma}}$$

$$= \frac{\tau_b.\tau_r}{(p_{0a}/p_a)^{(\gamma-1)/\gamma}}$$

$$= \frac{\tau_b.\tau_r}{\tau_r} = \tau_b$$

$$\frac{T_9}{T_a} = \tau_b$$

5. Enthalpy balance in combustion chamber

$$\dot{m}_a . h_{02} + \dot{m}_f h_{PR} = (\dot{m}_a + \dot{m}_f) h_{04}$$

As $\dot{m}_f$ is very small and neglected and $C_{p2} = C_{p4} = C_p$

$$\dot{m}_a C_p T_{02} + \dot{m}_f h_{PR} = \dot{m}_a C_p T_{04}$$

$$\dot{m}_f h_{PR} = \dot{m}_a C_p (T_{04} - T_{02})$$

$$= \dot{m}_a C_p T_{02} \left(\frac{T_{04}}{T_{02}} - 1\right)$$

The fuel/air ratio may be defined as,

$$f = \frac{\dot{m}_f}{\dot{m}_a} = \frac{C_p T_{02}}{h_{PR}} \left(\frac{T_{04}}{T_{02}} - 1\right)$$

For an ideal engine,

$$T_{0a} = T_{02} = T_a . \tau_r$$

$$\frac{T_{04}}{T_{02}} = \tau_b$$

therefore,

$$f = \frac{C_p T_a \tau_r}{h_{PR}} (\tau_b - 1)$$

now,

$$\tau_\lambda = \frac{C_p T_{04}}{C_p T_a} = \frac{T_{04}}{T_a}$$

$$= \frac{T_{02}}{T_a} . \frac{T_{04}}{T_{02}} = \frac{T_{0a}}{T_a} . \frac{T_{04}}{T_{02}}$$

$$= \tau_r . \tau_b$$

therefore,

$$f = \frac{C_p T_a}{h_{PR}} (\tau_\lambda - \tau_r)$$

6. In ramjet, there is no compressor and therefore no turbine.
7.

Since,

$$M_9 = M_a$$

and $\frac{T_9}{T_a} = \tau_b$

therefore,

$$\left(\frac{V_9}{a_a}\right)^2 = \frac{T_9}{T_a}M_9^2$$

$$= \tau_b . M_a^2$$

and thrust F,

$$F = \dot{m}_a . a_a \left(\frac{V_9}{a_a} - M_a\right)$$

$$= \dot{m}_a . a_a\left(\sqrt{\tau_b} . M_a - M_a\right)$$

$$= \dot{m}_a . a_a M_a\left(\sqrt{\tau_b} - 1\right)$$

$$= \dot{m}_a . a_a M_a\left(\sqrt{\frac{\tau_\lambda}{\tau_r}} - 1\right)$$

$$\frac{F}{\dot{m}_a} = a_a M_a\left(\sqrt{\frac{\tau_\lambda}{\tau_r}} - 1\right)$$

8. Thrust specific fuel consumption, S

$$S = \frac{f}{F/\dot{m}_a}$$

$$= \frac{C_p T_a(\tau_\lambda - \tau_r)}{h_{PR} . a_a . M_a\left(\sqrt{\tau_\lambda/\tau_r} - 1\right)}$$

$$= \frac{C_p T_a(\tau_r . \tau_b - \tau_r)}{h_{PR} . a_a . M_a\left(\sqrt{\tau_\lambda/\tau_r} - 1\right)}$$

$$= \frac{C_p T_a . \tau_r(\tau_b - 1)}{h_{PR} . a_a . M_a\left(\sqrt{\tau_b} - 1\right)}$$

9. Thermal efficiency, η_T

$$\eta_T = \frac{\text{Heat supplied} - \text{Heat rejected}}{\text{Heat supplied}}$$

$$= 1 - \frac{\text{Heat rejected}}{\text{Heat supplied}}$$

$$= 1 - \frac{C_p(T_9 - T_a)}{C_p(T_{04} - T_{02})}$$

$$= 1 - \frac{T_a\left(\frac{T_9}{T_a} - 1\right)}{T_{02}\left(\frac{T_{04}}{T_{02}} - 1\right)}$$

$$= 1 - \frac{T_a}{T_{02}}$$

$$\left(\text{as } \frac{T_{09}}{T_9} = \frac{T_{04}}{T_9} = \frac{T_{02}}{T_a} = \frac{T_{0a}}{T_a} = (\pi)^{(\gamma-1)/\gamma}\right)$$

$$\eta_T = 1 - \frac{T_a}{T_{0a}}$$

$$= 1 - \frac{1}{(T_{0a}/T_a)}$$

$$\eta_T = 1 - \frac{1}{\tau_r}$$

10. Propulsive efficiency, η_p

$$\eta_p = \frac{\text{useful thrust power}}{\text{change in kinetic energy}}$$

$$= \frac{\dot{m}_a V_a (V_9 - V_a)}{\frac{1}{2}\dot{m}_a\left({V_9}^2 - {V_a}^2\right)}$$

$$= \frac{2V_a}{V_9 + V_a}$$

$$= \frac{2}{\frac{V_9}{V_a} + 1}$$

For ramjet

$$M_a = M_9$$

$$\frac{V_a}{a_a} = \frac{V_9}{a_9}$$

$$a_9^2 = \gamma R T_9, a_a^2 = \gamma R T_a$$

$$\frac{V_a}{\sqrt{T_a}} = \frac{V_9}{\sqrt{T_9}}$$

$$\frac{V_9}{V_a} = \sqrt{\frac{T_9}{T_a}}$$

$$\tau_\lambda = \frac{T_{04}}{T_a} = \frac{T_{04}}{T_{02}} \cdot \frac{T_{02}}{T_a}$$

$$= \tau_b . \tau_r, \text{(as } T_{02} = T_{0a})$$

$$\frac{T_{09}}{T_9} = \frac{T_{04}}{T_9} = \frac{T_{0a}}{T_a}$$

$$\frac{T_9}{T_a} = \frac{T_{04}}{T_{0a}} = \frac{T_{04}}{T_{02}} = \tau_b$$

$$= \frac{\tau_\lambda}{\tau_r}$$

therefore,

$$\eta_P = \frac{2}{\sqrt{\frac{\tau_\lambda}{\tau_r}} + 1}$$

11. Overall efficiency

$$\eta_0 = \eta_P . \eta_T$$

EXAMPLE

Example 9.1

To understand the optimized Mach number for best performance of a ramjet engine, calculate the performance of a ramjet engine i.e., thrust/mass flow rate of air intake $(F/\dot{m}_a)$ for various Mach numbers $M_a = 1.5, 2, 2.5, 3$ and 3.5. The input data are:

$$T_a = 216.7\ K, \gamma = 1.4, C_p = 1.005\ kJ/kg.K, h_{PR} = 42{,}800\ kJ/kg,$$
$$T_{04} = 1600, 1900 \text{ and } 2200\ K$$

(*Source:* Mattingly, Jack D., Elements of Gas Turbine Propulsion, McGraw Hill Education, 2017)

Solution

Take $M_a = 2$ and $T_{04} = 1900\ K$

$$R = \frac{\gamma - 1}{\gamma} C_p$$

$$= \frac{1.4 - 1}{1.4} \times 1.005 \times 10^3$$

$$R = 287\ J/kg\ K$$

$$a_a = \sqrt{\gamma R T_a} = \sqrt{1.4 \times 287 \times 216.7}$$

$$a_a = 295\, m/s$$

$$\tau_r = 1 + \frac{\gamma - 1}{2}$$

$$M_a^2 = 1 + (0.2 \times 2^2)$$

$$\tau_r = 1.8$$

$$\tau_\lambda = \frac{T_{04}}{T_a} = \frac{1900}{216.7}$$

$$\tau_\lambda = 8.77$$

$$\frac{V_9}{a_a} = M_a\sqrt{\frac{\tau_\lambda}{\tau_r}} = 2\sqrt{\frac{8.77}{1.8}}$$

$$\frac{V_9}{a_a} = 4.414$$

$$\frac{F}{\dot{m}_a} = a_a M_a\left(\sqrt{\frac{\tau_\lambda}{\tau_r}} - 1\right)$$

$$= 295 \times 2\left(\sqrt{\frac{8.77}{1.8}} - 1\right)$$

$$\frac{F}{\dot{m}_a} = 712.313\, N.s/kg$$

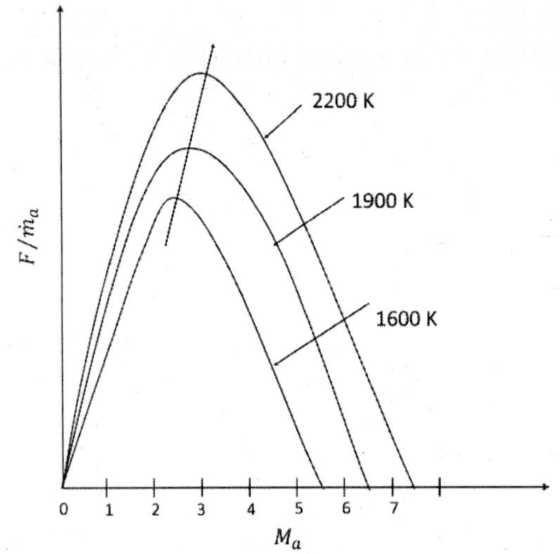

Figure 9.4: Ramjet Performance for the Given Example

Similarly, $F/\dot{m}_a$ can be calculated for all other Mach numbers and temperatures as given above and a plot showing the trend similar to Figure 9.4 can be drawn.

It may be noted that thrust per unit mass flow rate, $(F/\dot{m}_a)$ is maximum in all cases at about $M_a = 2.5$ and therefore, the ramjet performance is best at Mach number 2–3.

9.3 IDEAL TURBOJET

Figures 9.5 and 9.6 give the sketch with station numbering for an ideal turbojet engine and T-s diagram for the same. In turbojet, a compressor-turbine unit is added. The thrust of a ramjet tends to zero as the Mach number goes to zero. Thermal efficiency of a ramjet is also zero at Mach zero. In an ideal turbojet, a compressor with pressure ratio of 10 will give a thermal efficiency of about 50 percent at Mach zero.

In an ideal turbojet, the work done by turbine is equal to the work required by the compressor. The work, $\dot{W}_t = \dot{W}_c$ represents per unit time. The rest of the thermal energy available in the hot gases passes through nozzle to produce thrust.

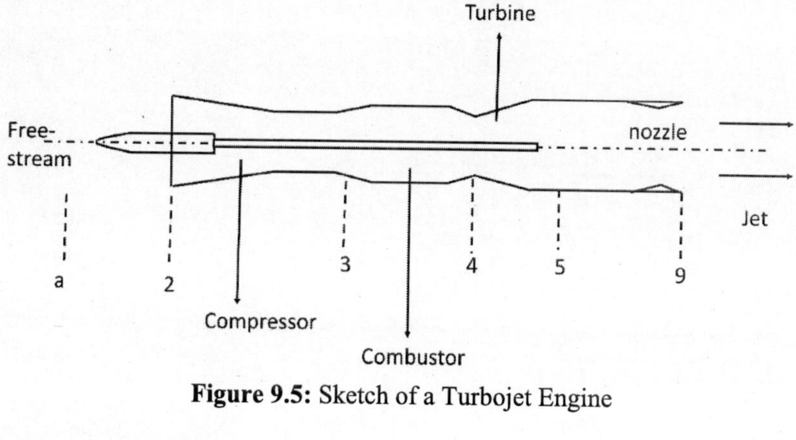

Figure 9.5: Sketch of a Turbojet Engine

Steps for cycle analysis

1. Uninstalled engine thrust, *F*

$$\frac{F}{\dot{m}_a} = (V_9 - V_a)$$

$$\frac{F}{\dot{m}_a} = a_a\left(\frac{V_9}{a_a} - M_a\right)$$

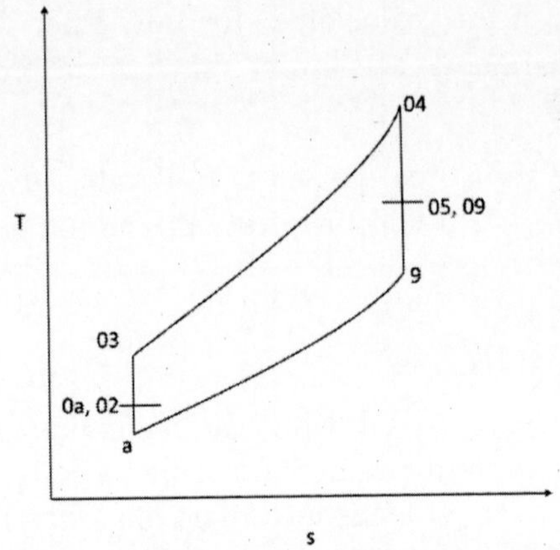

Figure 9.6: T-s Diagram of an Ideal Turbojet Engine
(*Source:* Mattingly, Jack D., Elements of Gas Turbine Propulsion, McGraw Hill Education, 2017)

2.

$$\left(\frac{V_9}{a_a}\right)^2 = \frac{T_9}{T_a} M_9^2$$

3.

$$\frac{p_{09}}{p_9} = \left(1 + \frac{\gamma - 1}{2} M_9^2\right)^{\gamma/(\gamma-1)}$$

$$p_{09} = p_a.\frac{p_{0a}}{p_a}.\frac{p_{02}}{p_{0a}}.\frac{p_{03}}{p_{02}}.\frac{p_{04}}{p_{03}}.\frac{p_{05}}{p_{04}}.\frac{p_{09}}{p_{05}}$$

$$= p_a.\pi_r.\pi_d.\pi_c.\pi_b.\pi_t.\pi_n$$

For an ideal engine,

$\pi_d = \pi_b = \pi_n = 1$

thus, $p_{09} = p_a.\pi_r.\pi_c.\pi_t$

$$M_9^2 = \frac{2}{(\gamma - 1)}\left[\left(\frac{p_{09}}{p_9}\right)^{(\gamma-1)/\gamma} - 1\right]$$

also,

$$\frac{p_{09}}{p_9} = \frac{p_{09}}{p_a}.\frac{p_a}{p_9}$$

$$= \pi_r.\pi_c.\pi_t.\frac{p_a}{p_9}$$

$(as, p_a = p_9)$

$$\frac{p_{09}}{p_9} = \pi_r.\pi_c.\pi_t$$

therefore,

$$M_9^2 = \frac{2}{(\gamma-1)}\left[(\pi_r.\pi_c.\pi_t)^{(\gamma-1)/\gamma} - 1\right]$$

as,

$$\tau_r = (\pi_r)^{(\gamma-1)/\gamma}$$

$$\tau_c = (\pi_c)^{(\gamma-1)/\gamma}$$

$$\tau_t = (\pi_t)^{(\gamma-1)/\gamma}$$

therefore,

$$M_9^2 = \frac{2}{(\gamma-1)}[(\tau_r.\tau_c.\tau_t) - 1]$$

4.

$$T_{09} = T_a.\frac{T_{0a}}{T_a}.\frac{T_{02}}{T_{0a}}.\frac{T_{03}}{T_{02}}.\frac{T_{04}}{T_{03}}.\frac{T_{05}}{T_{04}}.\frac{T_{09}}{T_{05}}$$

$$= T_a.\tau_r.\tau_d.\tau_c.\tau_b.\tau_t.\tau_n$$

$$\frac{T_9}{T_a} = \frac{(T_{09}/T_a)}{(T_{09}/T_9)}$$

$$= \frac{\tau_r.\tau_c.\tau_b.\tau_t}{(p_{09}/p_9)^{(\gamma-1)/\gamma}}$$

$$= \frac{\tau_r.\tau_c.\tau_b.\tau_t}{(\pi_r.\pi_c.\pi_t)^{(\gamma-1)/\gamma}}$$

$$= \frac{\tau_r.\tau_c.\tau_b.\tau_t}{\tau_r.\tau_c.\tau_t}$$

$$\frac{T_9}{T_a} = \tau_b$$

5. Enthalpy balance in combustion chamber

$$\dot{m}_a h_{03} + \dot{m}_f h_{pR} = (\dot{m}_a + \dot{m}_f)h_{04}$$

For an ideal cycle, $\dot{m}_f$ is very small and hence neglected.

also,

$$c_{p_3} = c_{P_4} = c_P$$

$$\dot{m}_a c_p T_{03} + \dot{m}_f h_{pR} = \dot{m}_a c_p T_{04}$$

$$\dot{m}_f h_{PR} = \dot{m}_a c_p (T_{04} - T_{03})$$

$$= \dot{m}_a \cdot c_p \cdot T_a \left(\frac{T_{04}}{T_a} - \frac{T_{03}}{T_a}\right)$$

by definition,

$$\tau_\lambda = \frac{T_{04}}{T_a}$$

$$\tau_\lambda = \frac{T_{04}}{T_{03}} . \frac{T_{03}}{T_{02}} . \frac{T_{02}}{T_a}$$

(as $T_{02} = T_{0a}$)

$$\tau_\lambda = \tau_b . \tau_c . \tau_r$$

and,

$$\frac{T_{03}}{T_a} = \frac{T_{03}}{T_{02}} . \frac{T_{02}}{T_a}$$

$$= \frac{T_{03}}{T_{02}} . \frac{T_{0a}}{T_a}$$

$$= \tau_c . \tau_r$$

therefore, fuel/air ratio f will be

$$f = \frac{\dot{m}_f}{\dot{m}_a}$$

$$= \frac{c_p T_a}{h_{PR}} . (\tau_\lambda - \tau_r . \tau_c)$$

$$= \frac{c_p T_a}{h_{PR}} . \tau_r \tau_c \left(\frac{\tau_\lambda}{\tau_r \tau_c} - 1\right)$$

$$= \frac{c_p T_a}{h_{PR}} . \tau_r \tau_c (\tau_b - 1)$$

6. The power output of the turbine per unit time,

$$\dot{W}_t = (\dot{m}_a + \dot{m}_f)(h_{04} - h_{05})$$

$$= \dot{m}_a C_p (T_{04} - T_{05})$$

$$= \dot{m}_a C_p T_{04} (1 - \tau_t)$$

(as, $\dot{m}_f$ is very small therefore neglected)

The power required by the compressor per unit time,

$$\dot{W}_c = \dot{m}_a(h_{03} - h_{02})$$

$$= \dot{m}_a C_p(T_{03} - T_{02})$$

$$= \dot{m}_a C_p T_{02}(\tau_c - 1)$$

For an ideal turbojet,

$$\dot{W}_c = \dot{W}_t$$

$$\dot{m}_a C_p T_{02}(\tau_c - 1) = \dot{m}_a C_p T_{04}(1 - \tau_t)$$

therefore,

$$\tau_t = 1 - \frac{T_{02}}{T_{04}}(\tau_c - 1)$$

$$= 1 - \frac{T_{02}}{T_a}\cdot\frac{T_a}{T_{04}}(\tau_c - 1)$$

$$= 1 - \frac{T_{0a}}{T_a}\cdot\frac{T_a}{T_{04}}(\tau_c - 1)$$

$$= 1 - \frac{\tau_r}{\tau_\lambda}(\tau_c - 1)$$

7.

$$\left(\frac{V_9}{a_a}\right)^2 = \frac{T_9}{T_a}M_9^2$$

$$= \tau_b \frac{2}{(\gamma - 1)}(\tau_r.\tau_c.\tau_t - 1)$$

$$= \frac{T_{04}}{T_a}\cdot\frac{T_a}{T_{02}}\cdot\frac{T_{02}}{T_{03}}\cdot\frac{2}{(\gamma - 1)}(\tau_r.\tau_c.\tau_t - 1)$$

$$= \frac{\tau_\lambda}{\tau_r\tau_c}\frac{2}{(\gamma - 1)}(\tau_r.\tau_c.\tau_t - 1)$$

8. Thrust specific fuel consumption

$$S = \frac{f}{F/\dot{m}_a}$$

9. Thermal efficiency, η_T

$$\eta_T = 1 - \frac{\text{Heat rejected}}{\text{Heat supplied}}$$

$$= 1 - \frac{C_p(T_9 - T_a)}{C_p(T_{04} - T_{03})} = 1 - \frac{T_a}{T_{03}}\frac{(T_9/T_a - 1)}{(T_{04}/T_{03} - 1)}$$

also,

$$\frac{T_{04}}{T_9} = \frac{T_{03}}{T_a} = (\pi)^{(\gamma-1)/\gamma}$$

$$\frac{T_{04}}{T_{03}} = \frac{T_9}{T_a}$$

$$\eta_T = 1 - \frac{T_a}{T_{03}}$$

$$= 1 - \frac{T_a}{T_{02}} \cdot \frac{T_{02}}{T_{03}}$$

$$(as, T_{02} = T_{0a})$$

$$= 1 - \frac{1}{\tau_r . \tau_c}$$

10. Propulsive efficiency, η_P

$$\eta_P = \frac{2V_a}{V_9 + V_a}$$

$$= \frac{2(V_a/a_a)}{\frac{V_9}{a_a} + \frac{V_a}{a_a}}$$

$$= \frac{2M_a}{(V_9/a_a) + M_a}$$

11. Overall efficiency, η_o

$$\eta_0 = \eta_P . \eta_T$$

EXAMPLE

Example 9.2

Calculate the performance parameters, $F/\dot{m}_a$, s, η_T and η_P for an ideal turbojet engine having the following input data:

$T_a = 216.7\ K$, $M_a = 2.0$, $\gamma = 1.4$, $C_p = 1.005\ kJ/kg.K$, $T_{04} = 1600\ K$, compressor pressure ratio, $\pi_c = 12$, $h_{PR} = 42{,}800\ kJ/kg$

Solution

$$R = \frac{\gamma - 1}{\gamma}$$

$$C_p = \frac{1.4 - 1}{1.4} \times 1.005$$

$$= 0.287\ kJ/kg.K$$

$$a_a = \sqrt{\gamma R T_a} = \sqrt{1.4 \times 0.287 \times 216.7 \times 1000}$$
$$= 295\, m/s$$

$$\tau_r = 1 + \frac{\gamma - 1}{2} M_a^2$$
$$= 1 + 0.2 \times 4$$
$$= 1.8$$

$$\tau_\lambda = \frac{T_{04}}{T_a} = \frac{1600}{216.7}$$
$$= 7.383$$

$$\tau_c = (\pi_c)^{(\gamma-1)/\gamma}$$
$$= (12)^{0.286}$$
$$= 2.035$$

$$\tau_t = 1 - \frac{\tau_r}{\tau_\lambda}(\tau_c - 1)$$
$$= 1 - \frac{1.8}{7.383}(2.035 - 1)$$
$$= 1 - 0.252$$
$$= 0.748$$

$$\frac{V_9}{a_a} = \sqrt{\frac{2}{(\gamma - 1)} . \frac{\tau_\lambda}{\tau_r . \tau_c}(\tau_r . \tau_c . \tau_t - 1)}$$
$$= \sqrt{\frac{2}{0.4} . \frac{7.383}{1.8 \times 2.035}(1.8 \times 2.035 \times 0.748 - 1)}$$
$$= 4.18$$

$$\frac{F}{\dot{m}_a} = a_a \left(\frac{V_9}{a_a} - M_a\right)$$
$$= 295(4.18 - 2.0)$$
$$= 643.1\, N/(kg/s) = 643.1\, N.s/kg$$

$$f = \frac{C_p T_a}{h_{PR}}(\tau_\lambda - \tau_r \tau_c)$$
$$= \frac{1.005 \times 216.7}{42800}(7.383 - 1.8 \times 2.035)$$
$$= 0.0189$$

$$s = \frac{f}{F/\dot{m}_a}$$
$$= \frac{0.0189}{643.1}$$
$$= 0.0000294 = 29.4\, mg/N.s$$

$$\eta_T = 1 - \frac{1}{\tau_r . \tau_c}$$
$$= 1 - \frac{1}{1.8 \times 2.035}$$
$$= 0.727 = 72.7\%$$

$$\eta_P = \frac{2M_a}{V_9/a_a + M_a}$$
$$= \frac{2 \times 2}{4.18 + 2} = \frac{4}{6.18}$$
$$= 0.647$$
$$= 64.7\%$$

9.4 IDEAL TURBOJET WITH AFTERBURNER

Afterburning involves burning additional fuel in the exhaust jet pipe. Thrust augmentation is required for a temporary increase in thrust during take-off, acceleration from subsonic to supersonic speed or during combat operations. Figures 9.7 and 9.8 give the schematic diagram with station numbering and T-s diagram for an ideal afterburning turbojet engine.

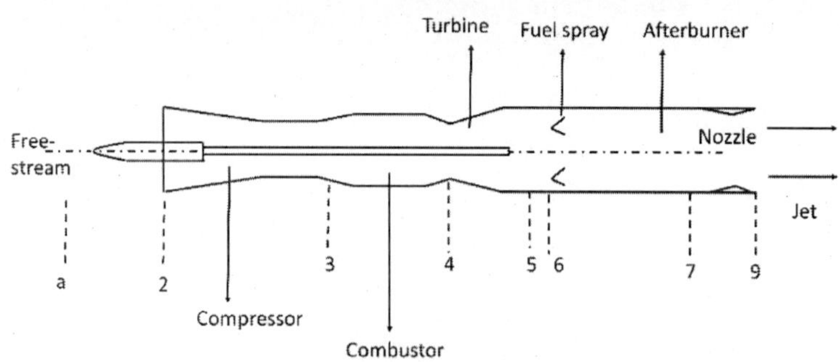

Figure 9.7: Sketch of a Turbojet Engine with Afterburner

The effect of additional fuel required in the afterburner on thrust augmentation can be estimated assuming the following operating parameters. Assume that the choked convergent nozzle is used, then the jet velocity will correspond to the sonic velocity at the appropriate temperature in the plane of the nozzle i.e., T_9 or $T_{9'}$ depending on whether the engine is operated with or without afterburner. T_{05} to $T_{9'}$ is the expansion in the nozzle without afterburner. T_{05} to T_{09} is the increase in temperature with additional fuel in the afterburner and T_{09} to T_9 is the expansion in the nozzle with afterburner.

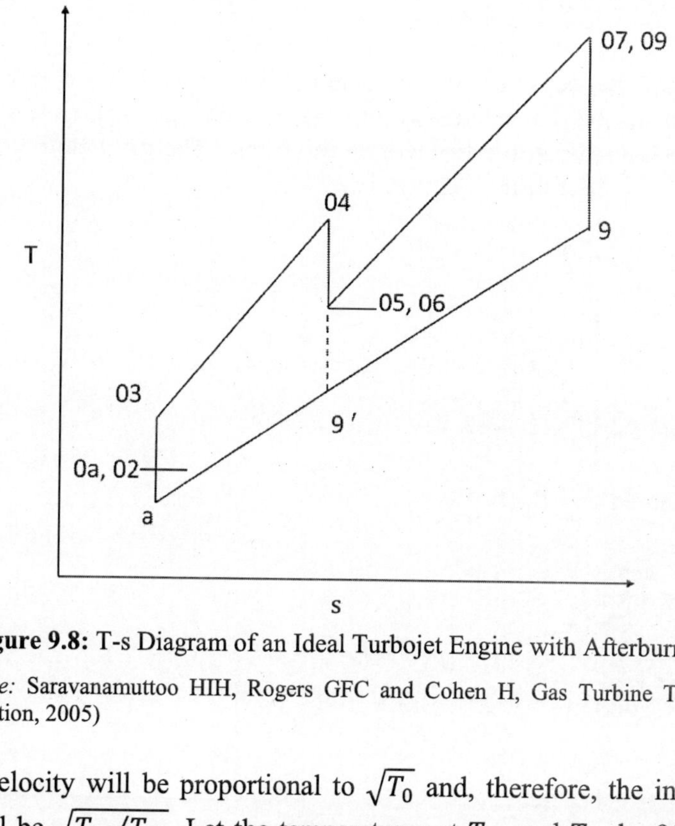

Figure 9.8: T-s Diagram of an Ideal Turbojet Engine with Afterburner

(*Source:* Saravanamuttoo HIH, Rogers GFC and Cohen H, Gas Turbine Theory, 5th Edition, 2005)

The jet velocity will be proportional to $\sqrt{T_0}$ and, therefore, the increase in thrust will be $\sqrt{T_{09}/T_{05}}$. Let the temperatures at T_{05} and T_{09} be 959 K and 2000 K respectively. Then the thrust will be increased by the factor $\sqrt{2000/959} = 1.44$. The corresponding increase in the fuel if the primary combustor operates between 565 K (compressor exit temperature) to 1200 K (temp. at the end of combustor) i.e., Temperatures T_{03} and T_{04} respectively and the expansion in turbine is from temperature T_{04}, 1200 K to 959 K, temperature T_{05}. The afterburner temperature increase is from 959 K to 2000 K, the temperature at T_{09}. The increase in fuel will be by a factor $[(2000 - 959) + (1200 - 565)]/(1200 - 565) = 2.64$. This shows that for 44% increase in thrust, 164% extra fuel will be required. Which means that afterburning should be used for short periods.

Afterburning offers greater gains for low by-pass ratio turbofans, because of the relatively low temperature after mixing of the hot and cold streams and the large quantities of excess air available for combustion. Military turbofans use afterburning for take-off and combat maneuvering.

9.4.1 Cycle Analysis

Steps

In addition to the steps involved in ideal turbojet analysis, the following extra steps need to be considered when the second combustion chamber as afterburner is incorporated to increase the thrust. The gas velocity at exit of the nozzle with afterburner is given by,

$$\frac{V_9^2}{2C_p} = T_{09} - T_9$$

$$= T_{09}\left[1 - \frac{T_9}{T_{09}}\right]$$

$$= T_{09}\left[1 - \left(\frac{p_9}{p_{09}}\right)^{(\gamma-1)/\gamma}\right]$$

It may be noted that T_{05} to T_{09} the combustion process is at constant pressure. Therefore,

$$\frac{p_9}{p_{09}} = \left(\frac{p_{9'}}{p_{05}}\right)_{dry}$$

$$= \left(\frac{p_9}{p_{09}}\right)_{AB}$$

Subscript:

- $AB -$ afterburner
- $dry -$ no afterburning

therefore,

$$\frac{V_{9'}^2}{2C_p} = T_{05} - T_{9'}$$

$$= T_{05}\left[1 - \frac{T_{9'}}{T_{05}}\right]$$

$$= T_{05}\left[1 - \left(\frac{p_{9'}}{p_{05}}\right)^{\frac{\gamma-1}{\gamma}}\right]$$

therefore,

$$\left(\frac{V_9}{V_{9'}}\right)^2 = \frac{T_{09}}{T_{05}}$$

and,

$$\left(\frac{V_9}{a_a}\right)^2 = \frac{T_{09}}{T_{05}}\left(\frac{V_{9'}}{a_a}\right)^2$$

The temperature ratio, $\tau_{\lambda\, AB}$ may be defined as

$$\tau_{\lambda\, AB} = \frac{T_{07}}{T_a} = \frac{T_{09}}{T_a}$$

$$[T_{09} = T_{07}]$$

then,

$$\frac{T_{09}}{T_{05}} = \frac{(T_{09}/T_{07})(T_{07}/T_a)}{(T_{05}/T_{04})(T_{04}/T_a)}$$

$$= \frac{\tau_{\lambda\, AB}}{\tau_t . \tau_\lambda}$$

therefore,

$$\left(\frac{V_9}{a_a}\right)^2 = \frac{2}{(\gamma-1)} . \frac{\tau_{\lambda\, AB}}{\tau_t . \tau_\lambda} . \frac{\tau_\lambda}{\tau_r . \tau_c}(\tau_r . \tau_c . \tau_t - 1)$$

$$= \frac{2}{(\gamma-1)} . \tau_{\lambda\, AB}\left(1 - \frac{1}{\tau_r . \tau_c . \tau_t}\right)$$

The total fuel input, $\dot{m}_f$ in both primary combustion chamber and afterburner is added which results in increase of temperature from T_{03} to T_{09},

$$\dot{m}_{f_{\text{Total}}} . h_{PR} = \dot{m}_a\, C_p\, (T_{09} - T_{03})$$

$$= \dot{m}_a C_p T_a \left(\frac{T_{09}}{T_a} - \frac{T_{03}}{T_a}\right)$$

$$= \dot{m}_a C_p T_a \left(\frac{T_{09}}{T_a} - \frac{T_{03}}{T_{02}} . \frac{T_{02}}{T_a}\right)$$

$$= \dot{m}_a C_p T_a\, (\tau_{\lambda\, AB} - \tau_c . \tau_r)$$

and total fuel air ratio,

$$f_{\text{total}} = \frac{\dot{m}_{f_{\text{Total}}}}{\dot{m}_a}$$

$$= \frac{C_p T_a}{h_{PR}}(\tau_{\lambda\, AB} - \tau_c . \tau_r)$$

and the thermal efficiency is given by,

$$\eta_T = \frac{\text{change in kinetic energy}}{\text{energy input}}$$

$$= \frac{\frac{1}{2}.\left[(\dot{m}_a + \dot{m}_{f\,\text{Total}})V_9^2 - \dot{m}_a V_a^2\right]}{\dot{m}_{f_{\text{Total}}}.h_{PR}}$$

$\dot{m}_{f\,\text{Total}} \ll \dot{m}_a$ and hence neglected

$$= \frac{\dot{m}_a[V_9^2 - V_a^2]}{2\,\dot{m}_{f_{\text{Total}}}.h_{PR}}$$

$$(as, a_a^2 = \gamma R T_a)$$

$$= (\gamma - 1)C_p T_a$$

$$\eta_T = \frac{(\gamma - 1)C_p T_a[(V_9/a_a)^2 - (M_a^2)]}{2\,f_{\text{Total}}.h_{PR}}$$

9.5 IDEAL TURBOFAN

The turbofan engine was developed to improve the propulsion efficiency of the jet engine by reducing the mean jet velocity. Lowering of jet velocity resulted in less engine noise. In a turbofan engine, a part of the total air intake by passes the core stream i.e., compressor, combustion chamber, turbine, and nozzle before being ejected through a separate propelling nozzle. The thrust is made of two components, the cold stream (or fan) thrust, and the hot stream (core stream) thrust. Sometimes it is desirable to mix the two streams and eject them as a single jet of reduced velocity.

Figures 9.9 and 9.10 give the schematic diagram with station numbering of a turbofan engine and T-s diagram for the same. The fan is also called a ducted propeller. The fan increases the inlet air mass flow rate. Both fan and compressor are driven by the power provided by the turbine.

The by-pass ratio, B is defined as the ratio of $\dot{m}_F$ mass flow rate of air to the fan or cold stream, to the mass flow rate of air entering the compressor or hot stream $\dot{m}_h$

$$B = \frac{\dot{m}_F}{\dot{m}_h}$$

and total mass flow rate, $\dot{m}_a$

$$\dot{m}_a = \dot{m}_F + \dot{m}_h$$

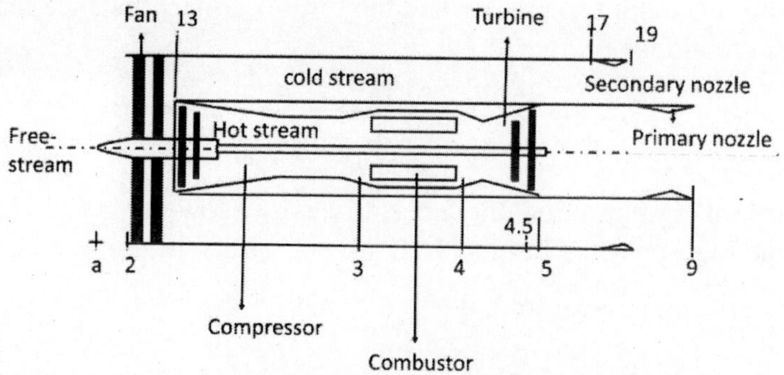

Figure 9.9: Turbofan Engine

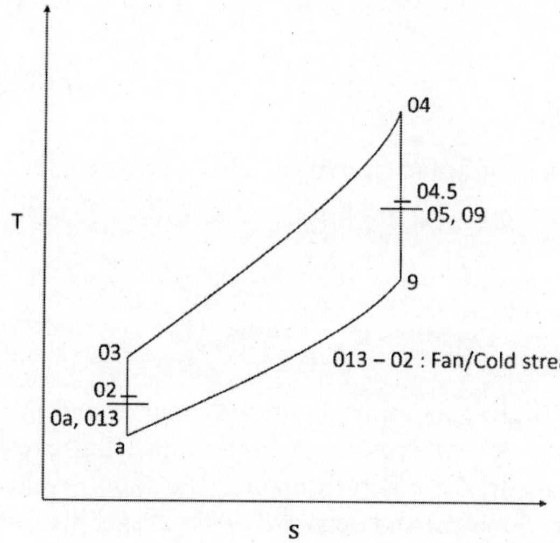

Figure 9.10: T-s Diagram of a Turbofan

therefore,

$$\dot{m}_F = \frac{\dot{m}_a B}{B+1}$$

$$\dot{m}_h = \frac{\dot{m}_a}{B+1}$$

When both the streams are ejected through the nozzle as separate streams at velocity, V_{jF} for fan/cold air and V_{jh} for products of combustion i.e., hot gases. The net thrust generated will be,

$$F = \left(\dot{m}_F . V_{jF} + \dot{m}_h . V_{jh}\right) - \dot{m}_a . V_a$$

where, V_a is the intake air velocity.

The turbine runs both compressor and fan. From enthalpy balance for turbine, compressor, and fan:

$$\dot{m}_h \, . C_p . (T_{04} - T_{04.5}) = \dot{m}_h \, . C_p . (T_{03} - T_{02})$$

$$\dot{m}_h . C_p . (T_{04.5} - T_{05}) = \dot{m}_F . C_p . (T_{02} - T_{013})$$

In a two-spool configuration, the fan is driven by a low-pressure turbine, LPT (4.5–5) and high-pressure turbine, HPT (4 – 4.5) runs the compressor.

The value of B may range from 0.3 to 8 or more.

9.5.1 Optimum Bypass Ratio

The bypass ratio, B is defined as,

$$B = \frac{\dot{m}_F}{\dot{m}_h}$$

$$\dot{m}_a = (1 + B)\dot{m}_h$$

The thrust for an ideal turbofan stream can also be written as,

$$F = \dot{m}_h(V_9 - V_a) + \dot{m}_F(V_{19} - V_a)$$

and,

$$\frac{F}{\dot{m}_a} = \frac{a_a}{(1 + B)}\left[\frac{V_9}{a_a} - M_a + B\left(\frac{V_{19}}{a_a} - M_a\right)\right]$$

For a given set of flight conditions T_a and M_a and design limit τ_λ, there are three design variables: compressor pressure ratio, fan pressure ratio and bypass ratio. Increasing the compressor pressure ratio results in increase in thrust per unit mass flow rate and decrease in thrust specific fuel consumption. Similarly, an optimum fan pressure ratio exists with all other parameters fixed. This optimum pressure ratio gives both, the maximum thrust per unit mass flow and minimum thrust specific fuel consumption. An optimum bypass ratio also exists which will give minimum thrust specific fuel consumption.

The thrust specific fuel consumption,

$$S = \frac{f}{(1 + B)(F/\dot{m}_a)}$$

Differentiate S with respect to B and put equal to zero to get optimum S.

$$\frac{\partial S}{\partial B} = 0$$

We get $(V_{19} - V_a) = 2(V_9 - V_a)$. This shows that when the bypass ratio is chosen to give the minimum specific fuel consumption, the thrust per unit mass flow of the engine core is one half that of the fan.

At the same overall pressure ratio, calculations may be made for a series of bypass ratio, by increasing the compressor pressure ratio. The plots may be drawn, and the following observations may be made:

1. Increasing bypass ratio improves SFC (specific fuel consumption) at the expense of a significant reduction in specific thrust.
2. The optimum fan pressure ratio increases with turbine inlet temperature.
3. The optimum fan pressure ratio decreases with the increase of bypass ratio.
4. The optimum SFC requires low specific thrust, this is important for high bypass turbofans.

9.5.2 Mixing of Hot and Fan Streams

Figure 9.11 gives the T-s diagram for mixing of hot and cold streams. The mixed stream will have temperature $06A$ and then gases will be expanded in jet to state $9'$.

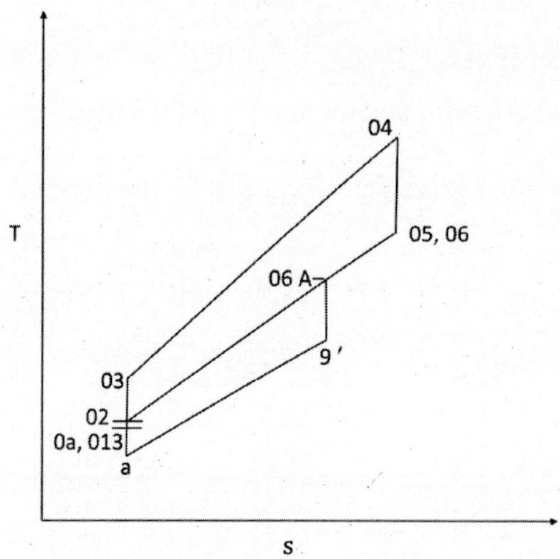

Figure 9.11: T-s Diagram for Mixed Hot and Cold Gases in a Turbofan $06A - 9'$ is the Propelling Nozzle.

Mixing is essential for an afterburning turbofan. The analysis is done assuming that mixing takes place in a constant area duct, with no losses and the flow is adiabatic.

From enthalpy balance.

$$\dot{m}_F . C_{p\,a} . T_{02} + \dot{m}_h . C_{p\,g} . T_{05} = \dot{m}_a . C_{p\,m} . T_{06\,A}$$

where $C_{p\,a}, C_{p\,g}$ and $C_{p\,m}$ are specific heats at constant pressure for fan/cold air, hot stream, and mixed stream respectively.

In ideal system,

$$C_{pa} = C_{pg} = C_{pm} = C_p$$

$$\dot{m}_F . T_{02} + \dot{m}_h . T_{05} = \dot{m}_a . T_{06\,A}$$

From momentum balance,

$$\dot{m}_F . V_F + \dot{m}_h . V_h = \dot{m}_a . V_{9'}$$

where, V_F is the velocity of the fan stream air, V_h is the velocity of the hot/core stream and $V_{9'}$ is the velocity of the mixed jet.

The continuity equation will also be required in analysis and is as follows:

$$\dot{m}_9 = \rho_9 A_9 V_9'$$

where ρ_9 is density of gases at station 9 and A_9 is the area of the nozzle.

9.6 IDEAL MIXED FLOW TURBOFAN WITH AFTERBURNER

The schematic diagram with station numbering is shown in Figure 9.12. The core flow, bypass flow and the mixed flow before passing through an afterburner and nozzle. Figure 9.13 gives the T-s diagram for the same.

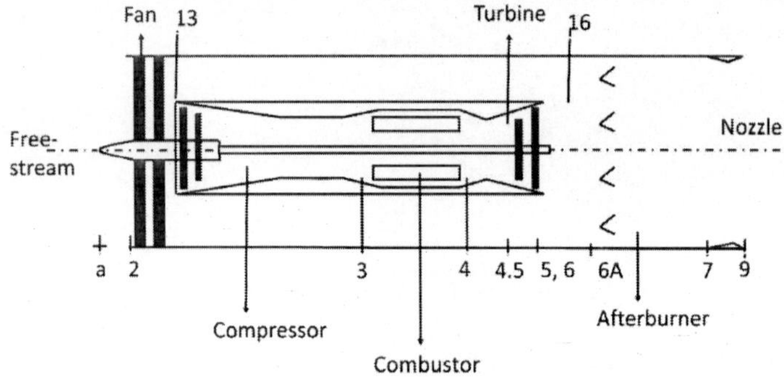

Figure 9.12: Station Numbering for Mixed Flow Turbofan with Afterburner

Since the engine has single inlet and exhaust, the ideal thrust of the engine is given by,

$$\frac{F}{\dot{m}_a} = a_a \left(\frac{V_9}{a_a} - M_a \right)$$

The analysis for the afterburner is similar to that for turbojet with afterburner.

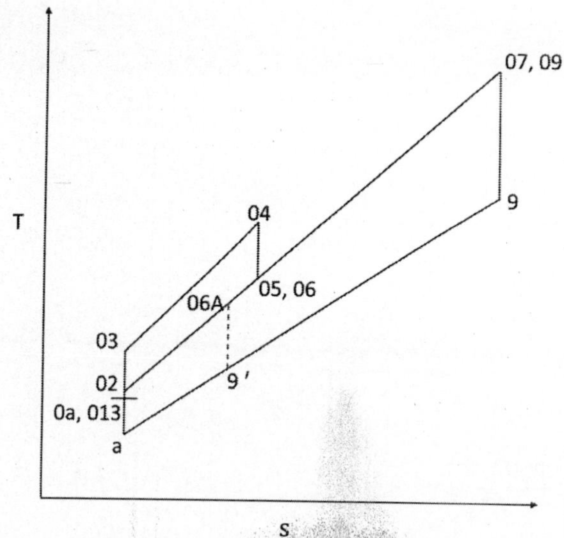

Figure 9.13: T-s Diagram for a Mixed Flow Turbofan with Afterburner
(*Source:* Mattingly, Jack D., Elements of Gas Turbine Propulsion, McGraw Hill Education, 2017)

9.7 IDEAL TURBOPROP

Figures 9.14 and 9.15 give the schematic diagram with station numbering of a turboprop engine and T-s diagram for an ideal cycle.

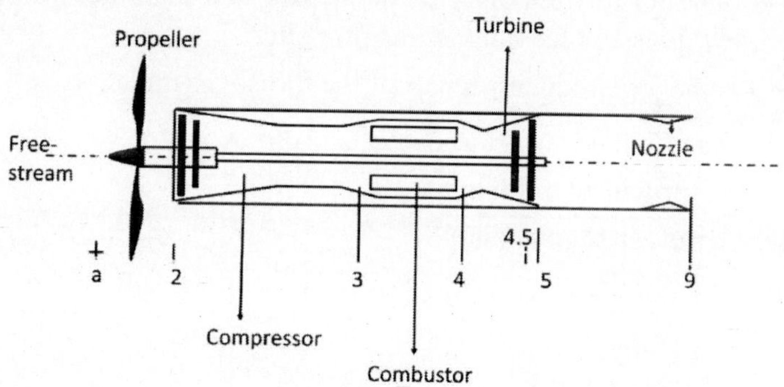

Figure 9.14: Station Numbering of Turboprop Engine

In a turboprop, the propeller tip Mach number will be very large at approximately M = 0.7 and above. The resultant loss in propeller efficiency limits the turboprop use up to Mach numbers less than 0.7. Presently, very high bypass ratio turbofans are preferred. Turboprops lead to very high noise levels, both in the airport vicinity and within the aircraft at flight speeds.

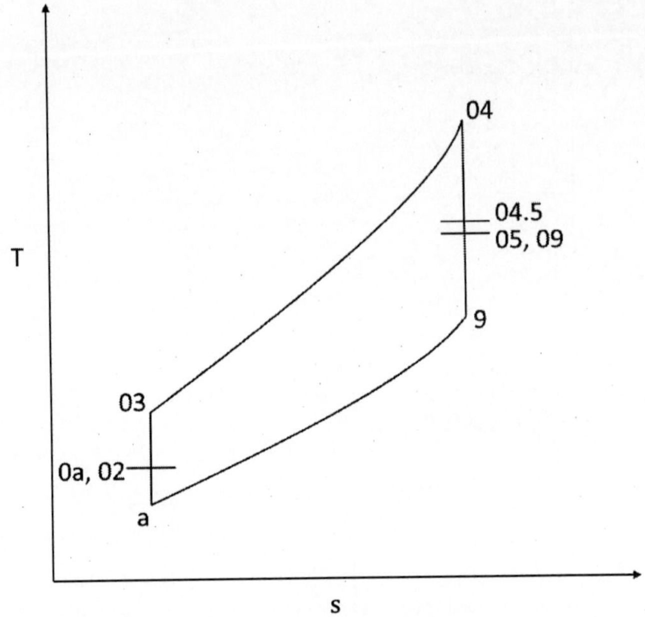

Figure 9.15: The T-s Diagram for a Turboprop Engine
(*Source:* Mattingly, Jack D., Elements of Gas Turbine Propulsion, McGraw Hill Education, 2017)

The turboprop has two turbines, i.e., high pressure, 04 – 04.5 in schematic and also in T-s diagram for running the compressor and the other a free turbine 04.5 – 05, low pressure for running the propeller.

- 09 – 9 is the jet, producing a part of the thrust generated.

The propulsive efficiency of a turboprop is given by

$$\eta_{prop} = \frac{\text{power to vehicle}}{\text{power to propeller}}$$

$$= \frac{F_{prop} . V_a}{\dot{W}_{prop}}$$

$$= \frac{\dot{m}_{prop}(V_j - V_a)}{\dot{W}_{prop}}$$

where, $\dot{m}_{prop}$ = mass of air passing through propeller

V_a = flight speed

V_j = jet speed

$\dot{W}_{prop}$ = propeller power in

The propeller efficiency increases with propeller size and the propeller size is limited by the onset of tip Mach number, losses, etc.

In a turboprop, the core stream exit nozzle is considered as unchoked.

In cycle analysis, for turboprop engine, the work supplied to the vehicle rather than the thrust is considered. A new dimensionless work output coefficient, C is defined as

$$C = \frac{\text{power interaction/mass flow of air through engine core}}{h_a}$$

where, h_a is the enthalpy of free stream air.

For the thrust of the core/hot stream, the work output coefficient is defined as

$$C_h = \frac{F_h . V_a}{\dot{m}_a . C_p . T_a}$$

Thus, the work output coefficient for the total turboprop engine is,

$$C_{tot} = C_{prop} + C_h$$

and the corresponding thrust is,

$$F = F_{prop} + F_h$$

$$= \frac{C_{tot} . \dot{m}_a . C_p . T_a}{V_a}$$

The uninstalled thrust F_h generated through engine core:

$$\frac{F_h}{\dot{m}_a} = a_a \left(\frac{V_9}{a_a} - M_a \right)$$

where, $\dot{m}_a$ is the mass of air passing through the engine core i.e., compressor, combustion chamber, both high pressure and low-pressure turbine and jet.

For the low-pressure turbine, also called free turbine, the power output is equal to the power input to the propeller,

$$\dot{W}_{prop} = \dot{m}_{4.5} . C_p . (T_{04.5} - T_{05})$$

also,

$$\dot{m}_{4.5} = \dot{m}_a + \dot{m}_f$$

$$= \dot{m}_a \text{ (as } \dot{m}_f \text{ is very small and neglected)}$$

$$C_{prop} = \frac{\eta_{prop} \dot{W}_{prop}}{\dot{m}_a C_p T_a}$$

The thermal efficiency of the engine cycle is the same as in the case of turbojet.

9.8 IDEAL TURBOSHAFT

In turboshaft engine, the output power drives a helicopter rotor. In helicopter applications, free turbine configuration is used. The helicopter rotor operates at a near constant speed by changing the pitch and changing the input energy to the combustion chamber. A helicopter flies at low altitude and at a lower speed. The speed is low due to aerodynamic limitations on the rotor blades. The jet thrust is not critical, and more shaft power is available.

The T-s diagram for a turboshaft engine is same as that for a turboprop engine given in figure 9.15.

SUMMARY

In this chapter, parametric cycle analysis i.e., performance of engines at different flight conditions are studied. The thrust and thrust specific fuel consumption are determined for a particular compressor pressure ratio, fan pressure ratio, bypass ratio etc. The design limitations are burner exit temperature, compressor exit pressure etc. and the flight conditions are Mach number, ambient temperature etc. The analysis is carried out in terms of two dimensionless parameters, π, the total pressure ratio at exit to entry of the component and τ, the total temperature ratio at exit to entry of the component.

The various engines analyzed include ideal ramjet, ideal turbojet, ideal turbojet with afterburner, ideal turbofan, ideal turbofan with afterburner, ideal turboprop, and ideal turboshaft. In most of the cases, the thermal efficiency, η_T and propulsive efficiency, η_P have also been estimated.

QUESTIONS

1. With the aid of a neat diagram, explain the working principle of a ramjet engine. Also, draw its thermodynamic cycle.
2. Derive the expression for thermal efficiency of an ideal turbojet engine.
3. Show that the expression for propulsive efficiency of an ideal ramjet engine is as given below:

$$\eta_p = \frac{2}{\left[\left(\frac{\tau_\lambda}{\tau_r}\right)^{1/2} + 1\right]}$$

4. Ramjet performance is best suited when it operates at about 2.5 Mach number. Explain.

CHAPTER 10

Real Jet Engine Cycle Analysis

10.1 INTRODUCTION

In real cycle analysis, the component efficiencies and losses will be taken into consideration. Both compression and expansion processes are irreversible adiabatic involving increase in entropy. Due to friction, there will be a pressure loss in the combustion chamber. The values of C_p and γ of the working fluid vary during the cycle. The fluid is air in compressor and products of combustion in turbine and nozzle. Transmission losses between turbine and compressor also need consideration.

10.2 TURBOJET

The sketch and station numbering of turbojet engine is given in Figure 9.5. One-dimensional flow of fluid at the entrance and exit of each component is assumed. The specific heat at constant pressure for upstream of combustor, which is air is C_{pa} and downstream of combustor is C_{pg}. The fluid is assumed to be perfect gas. Figure 10.1 gives the T-s diagram with losses.

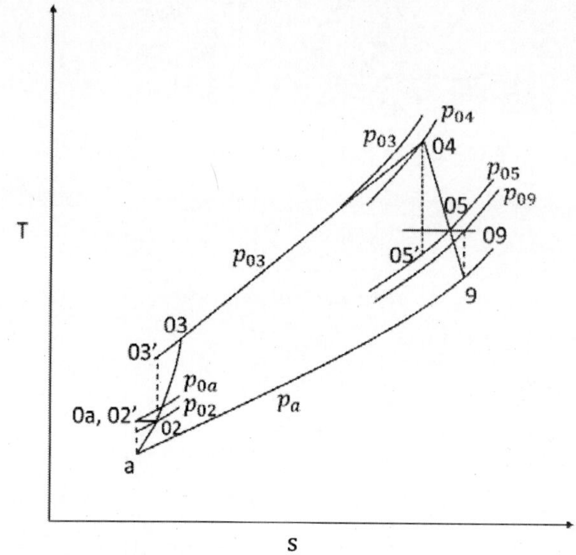

Figure 10.1: T-s Diagram with Losses for a Turbojet Engine

Parametric Cycle Analysis

Step 1

Uninstalled Thrust

$$F = (\dot{m}_9 V_9 - \dot{m}_a V_a) + A_9(p_9 - p_a)$$

$$\frac{F}{\dot{m}_a} = a_a\left(\frac{\dot{m}_9 V_9}{\dot{m}_a a_a} - M_a\right) + \frac{A_9 p_9}{\dot{m}_a}\left(1 - \frac{p_a}{p_9}\right)$$

Now

$$\frac{A_9 p_9}{\dot{m}_a}\left(1 - \frac{p_a}{p_9}\right) = \frac{\dot{m}_9}{\dot{m}_a} \cdot \frac{A_9 p_9}{\dot{m}_9}\left(1 - \frac{p_a}{p_9}\right)$$

$$= \frac{\dot{m}_9}{\dot{m}_a} \cdot \frac{A_9 p_9}{\rho_9 A_9 V_9}\left(1 - \frac{p_a}{p_9}\right)$$

(from continuity equation, $\dot{m}_9 = \rho_9 A_9 V_9$)

$$= \frac{\dot{m}_9}{\dot{m}_a} \cdot \frac{p_9}{(p_9/R_9 T_9)V_9}\left(1 - \frac{p_a}{p_9}\right)$$

(from equation of state, $p_9 = \rho_9 R_9 T_9$)

$$= \frac{\dot{m}_9}{\dot{m}_a} \cdot \frac{R_9 T_9}{V_9}\left(1 - \frac{p_a}{p_9}\right)$$

$$= \frac{\dot{m}_9}{\dot{m}_a} \cdot \frac{R_9 T_9}{V_9} \cdot \frac{\gamma_a R_a T_a}{\gamma_a R_a T_a}\left(1 - \frac{p_a}{p_9}\right)$$

$$= \frac{\dot{m}_9}{\dot{m}_a} \cdot \frac{R_9 T_9}{R_a T_a} \cdot \frac{a_a^2}{\gamma_a V_9}\left(1 - \frac{p_a}{p_9}\right)$$

(as, $a_a^2 = \gamma_a R_a T_a$)

therefore,

$$\frac{A_9 p_9}{\dot{m}_a}\left(1 - \frac{p_a}{p_9}\right) = a_a\left[\frac{\dot{m}_9 R_9\, T_9/T_a}{\dot{m}_a R_a\, V_9/a_a} \cdot \frac{(1 - p_a/p_9)}{\gamma_a}\right]$$

also,

$$\dot{m}_9 = \dot{m}_a + \dot{m}_f$$

$$\frac{\dot{m}_9}{\dot{m}_a} = 1 + \frac{\dot{m}_f}{\dot{m}_a}$$

$$= 1 + f$$

where, f is fuel air ratio.

therefore,

$$\frac{F}{\dot{m}_a} = a_a\left[(1+f)\frac{V_9}{a_a} - M_a + (1+f)\frac{R_g}{R_a}.\frac{T_9/T_a}{V_9/a_a}.\frac{(1-p_a/p_9)}{\gamma_a}\right]$$

$(R_9 = R_g$ for downstream of combustor $i.e.$ in turbine and nozzle)

Step 2

$$\left(\frac{V_9}{a_a}\right)^2 = \frac{a_9^2 M_9^2}{a_a^2}$$

$$= \frac{\gamma_9 R_9 T_9 M_9^2}{\gamma_a R_a T_a}$$

$$= \frac{\gamma_g R_g T_9 M_9^2}{\gamma_a R_a T_a}$$

where in compressor $\gamma_a R_a$ are properties of fluid, which is air and $\gamma_g R_g$ are the properties of fluid i.e., products of combustion passing through the turbine and nozzle.

by definition,

$$\tau_\lambda = \frac{C_{pg}.T_{04}}{C_{pa}.T_a}$$

Step 3

$$M_9^2 = \frac{2}{(\gamma_g - 1)}\left[\left(\frac{p_{09}}{p_9}\right)^{(\gamma_g-1)/\gamma_g} - 1\right]$$

$$\frac{p_{09}}{p_9} = \frac{p_a}{p_9}.\pi_r.\pi_d.\pi_c.\pi_b.\pi_t.\pi_n$$

Step 4

$$\frac{T_9}{T_a} = \frac{(T_{09}/T_a)}{(p_{09}/p_a)^{(\gamma_g-1)/\gamma_g}}$$

also,

$$\frac{T_{09}}{T_a} = \tau_r.\tau_d.\tau_c.\tau_b.\tau_t.\tau_n$$

Step 5

Enthalpy balance in combustion chamber

$$\dot{m}_a.C_{pa}.T_{03} + \eta_b.\dot{m}_f.h_{PR} = \dot{m}_4.C_{pg}.T_{04}$$

where, η_b is combustion efficiency in burner,

Therefore, by dividing enthalpy balance equation by $\dot{m}_a . C_{pa} . T_a$

$$\frac{T_{03}}{T_a} + \eta_b \frac{\dot{m}_f}{\dot{m}_a} . \frac{h_{PR}}{C_{pa} . T_a} = \frac{\dot{m}_4 . C_{pg} . T_{04}}{\dot{m}_a . C_{pa} . T_a}$$

$$\dot{m}_4 = \dot{m}_a + \dot{m}_f$$

$$f = \frac{\dot{m}_f}{\dot{m}_a}$$

therefore,

$$\frac{\dot{m}_4}{\dot{m}_a} = (1 + f)$$

$$\frac{T_{03}}{T_a} + f . \frac{\eta_b . h_{PR}}{C_{pa} . T_a} = (1 + f) . \tau_\lambda$$

since,

$$\frac{T_{03}}{T_a} = \frac{T_{03}}{T_{02}} . \frac{T_{02}}{T_a}$$

$$= \tau_c . \tau_r$$

therefore,

$$\tau_r . \tau_c + f . \frac{\eta_b . h_{PR}}{C_{pa} T_a} = (1 + f) \tau_\lambda$$

we can easily get,

$$f = \frac{\tau_\lambda - \tau_r . \tau_c}{\left(\eta_b . h_{PR} / C_{pa} . T_a\right) - \tau_\lambda}$$

Step 6: The power output of the turbine will be equal to the power required by the compressor with a mechanical efficiency, η_m for losses in coupling.

$$\dot{m}_a C_{pa} (T_{03} - T_{02}) = \eta_m . \dot{m}_4 . C_{pg} (T_{04} - T_{05})$$

Divide this by $\dot{m}_a . C_{pa} . T_{02}$

$$\frac{T_{03}}{T_{02}} - 1 = \eta_m . (1 + f) . \frac{C_{pg} . T_{04}}{C_{pa} . T_{02}} \left(1 - \frac{T_{05}}{T_{04}}\right)$$

$$\tau_c - 1 = \eta_m . (1 + f) . \frac{C_{pg} . T_{04}}{C_{pa} . T_a} . \frac{T_a}{T_{02}} (1 - \tau_t)$$

Assuming, $C_{pg} = C_{pa}$

$$\tau_c - 1 = \eta_m . (1 + f) . \frac{\tau_\lambda}{\tau_r} (1 - \tau_t)$$

The same equation can be solved in terms of turbine temperature ratio, τ_t as:

$$\tau_t = 1 - \frac{1}{\eta_m(1+f)} \cdot \frac{\tau_r}{\tau_\lambda}(\tau_c - 1)$$

Compressor efficiency, from Figure 10.1,

$$\eta_c = \frac{T'_{03} - T_{02}}{T_{03} - T_{02}}$$

$$= \frac{T_{02}\left(\frac{T'_{03}}{T_{02}} - 1\right)}{T_{02}\left(\frac{T_{03}}{T_{02}} - 1\right)} = \frac{\pi_c{}^{(\gamma_a-1)/\gamma_a} - 1}{(\tau_c - 1)}$$

as,

$$\frac{T'_{03}}{T_{02}} = \left(\frac{p_{03}}{p_{02}}\right)^{(\gamma_a-1)/\gamma_a}$$

$$= (\pi_c)^{(\gamma_a-1)/\gamma_a}$$

and

$$\frac{T_{03}}{T_{02}} = \tau_c$$

and turbine efficiency, Figure 10.1,

$$\eta_t = \frac{T_{04} - T_{05}}{T_{04} - T_{05'}}$$

$$= \frac{T_{04}\left(1 - \frac{T_{05}}{T_{04}}\right)}{T_{04}\left(1 - \frac{T_{05'}}{T_{04}}\right)}$$

also,

$$\frac{T_{05'}}{T_{04}} = \left(\frac{p_{05'}}{p_{04}}\right)^{(\gamma_g-1)/\gamma_g}$$

$$= (\pi_t)^{(\gamma_g-1)/\gamma_g}$$

$$\eta_t = \frac{1 - \tau_t}{1 - (\pi_t)^{(\gamma_g-1)/\gamma_g}}$$

The polytropic efficiency of a compressor may be defined as

$$e_c = \frac{\text{Ideal work of compression for a differential pressure change}}{\text{actual work of compression for a differential pressure change}}$$

As work is proportional to enthalpy and enthalpy is proportional to temperature,

$$e_c = \frac{dT_{0i}}{dT_0}$$

where i is a stage of compressor

For ideal compressor, the isentropic relation is,

$$T_{0i} = p_{0i}{}^{(\gamma-1)/\gamma} \times \text{constant}$$

take log and differentiate,

$$\frac{dT_{0i}}{T_{0i}} = \frac{(\gamma-1)}{\gamma} \cdot \frac{dp_{0i}}{p_{0i}}$$

This can be re-written as,

$$\frac{dT_{0i}}{T_{0i}} = \frac{(\gamma-1)}{\gamma} \cdot \frac{dp_0}{p_0}$$

$$e_c = \frac{dT_{0i}}{dT_0} = \frac{dT_{0i}/T_0}{dT_0/T_0}$$

$$= \frac{(\gamma-1)}{\gamma} \cdot \frac{(dp_0/p_0)}{(dT_0/T_0)}$$

Assuming that polytropic efficiency e_c is constant,

$$\frac{dT_0}{T_0} = \frac{(\gamma-1)}{\gamma . e_c} \cdot \left(\frac{dp_0}{p_0}\right)$$

Integrate between T_{02} to T_{03},

$$\log_e \frac{T_{03}}{T_{02}} = \frac{(\gamma-1)}{\gamma . e_c} . \log_e \left(\frac{p_{03}}{p_{02}}\right)$$

$$= \log_e \left(\frac{p_{03}}{p_{02}}\right)^{((\gamma-1)/\gamma . e_c)}$$

$$\log_e \tau_c = \log_e (\pi_c)^{((\gamma-1)/\gamma . e_c)}$$

on integrating the above, we get

$$\tau_c = (\pi_c)^{((\gamma-1)/\gamma . e_c)}$$

and,

$$\eta_c = \frac{(\pi_c)^{(\gamma_a-1)/\gamma_a} - 1}{(\pi_c)^{(\gamma_a-1)/(\gamma_a . e_c)} - 1}$$

For compressor, γ is replaced by γ_a.

Similarly, defining polytropic efficiency for turbine as e_t, the turbine efficiency will be

$$\eta_t = \frac{1-(\pi_t)^{((\gamma_g-1)/(\gamma_g.e_t))}}{1-(\pi_t)^{((\gamma_g-1)/\gamma_g)}}$$

Step 7: The specific thrust $F/\dot{m}_a$ can now be evaluated.

Step 8: The thrust specific fuel consumption,

$$S = \frac{f}{(F/\dot{m}_a)}$$

Step 9: Evaluation of propulsive efficiency, η_p

$$\eta_p = \frac{\text{Thrust power}}{\text{Change in kinetic energy}}$$

$$= \frac{F.V_a}{\frac{1}{2}(\dot{m}_9 V_9^2 - \dot{m}_a V_a^2)}$$

$$= \frac{2V_a(F/\dot{m}_a).\dot{m}_a}{\left((\dot{m}_a + \dot{m}_f)V_9^2 - \dot{m}_a V_a^2\right)}$$

$$= \frac{2.V_a(F/\dot{m}_a)}{(1+f)V_9^2 - V_a^2}$$

$$= \frac{2.V_a.(F/\dot{m}_a)}{a_a^2[(1+f)(V_9^2/a_a^2) - (V_a^2/a_a^2)]}$$

$$= \frac{2.V_a.(F/\dot{m}_a)}{a_a^2[(1+f)(V_9^2/a_a^2) - M_a^2]}$$

Step 10: Evaluation of thermal efficiency, η_T

$$\eta_T = \frac{\text{Change in kinetic energy}}{\text{heat released from fuel}}$$

$$= \frac{\frac{1}{2}[(\dot{m}_a + \dot{m}_f)V_9^2 - \dot{m}_a V_a^2]}{\dot{m}_f.h_{PR}}$$

$$= \frac{a_a^2[(1+f)\,V_9^2/a_a^2 - M_a^2]}{2.f.h_{PR}}$$

10.3 TURBOJET WITH AFTERBURNER

The schematic diagram with station numbering for a turbojet with afterburner is given in Figure 9.7. The T-s diagram with losses for the same is given in Figure 10.2.

With afterburner

$$\pi_{AB} = \frac{p_{07}}{p_{05}} \quad f_{AB} = \frac{\dot{m}_{f\,AB}}{\dot{m}_a}$$

$$\tau_{AB} = \frac{T_{07}}{T_{06}} = \frac{T_{07}}{T_{05}} \quad \tau_{\lambda\,AB} = \frac{C_{pg}}{C_{pa}}\cdot\frac{T_{07}}{T_a}$$

$$\eta_{AB} = \frac{(\dot{m}_a + \dot{m}_f + \dot{m}_{fAB})C_{pg}T_{07} - (\dot{m}_a + \dot{m}_f)C_{pg}.T_{06}}{\dot{m}_{f\,AB}h_{PR}}$$

Where η_{AB} is the burning efficiency of the afterburner and is defined as the ratio of heat added in the afterburner to the heat released from fuel in the afterburner. $\dot{m}_{f\,AB}$ is the fuel supplied to the afterburner. Station 05 is the exit of turbine and station 06 is the entry to the afterburner. The temperatures T_{05} and T_{06} are same.

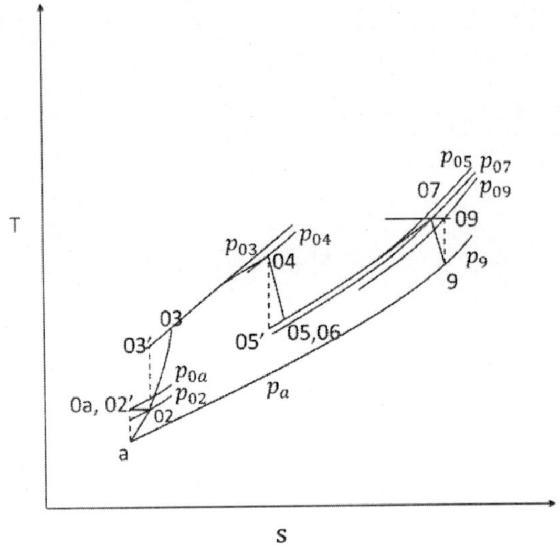

Figure 10.2: T-s Diagram with Losses for a Turbojet Engine with Afterburner

$$\tau_{\lambda\,AB} = \frac{C_{pg}}{C_{pa}}\cdot\frac{T_{04}}{T_a}\cdot\frac{T_{05}}{T_{04}}\cdot\frac{T_{07}}{T_{05}}$$

$$= \frac{C_{pg}}{C_{pa}}.\tau_{\lambda}.\tau_t.\tau_{AB}$$

and,

$$\frac{\dot{m}_9}{\dot{m}_a} = \frac{\dot{m}_a + \dot{m}_f + \dot{m}_{f\,AB}}{\dot{m}_a}$$
$$= 1 + f + f_{AB}$$

Where, $f_{AB} = \frac{\dot{m}_{f\,AB}}{\dot{m}_a}$

The fuel air ratio in the afterburner.

In cycle analysis, the effect of fuel addition in the afterburner should be included. An extra term f_{AB} will be introduced along with f.

Step 1

The uninstalled engine thrust,

$$\frac{F}{\dot{m}_a} = a_a\left[(1+f+f_{AB}).\frac{V_9}{a_a} - M_a + (1+f+f_{AB}).\frac{R_g}{R_a}.\left(\frac{T_9}{T_a}\Big/\frac{V_9}{a_a}\right).\frac{\left(1 - {p_a}/{p_9}\right)}{\gamma_a}\right]$$

where, R_g is the gas constant in afterburner.

Step 2

$$\left(\frac{V_9}{a_a}\right)^2 = \frac{\gamma_g\, R_g\, T_9\, M_9^2}{\gamma_a R_a T_a}$$

Here, γ and R terms are taken for afterburner.

Step 3

$$M_9^2 = \frac{2}{(\gamma_g - 1)}\left[\left(\frac{p_{09}}{p_9}\right)^{(\gamma_g-1)/\gamma_g} - 1\right]$$

$$\frac{p_{09}}{p_9} = \frac{p_a}{p_9}.\pi_r.\pi_d.\pi_c.\pi_b.\pi_t.\pi_{AB}.\pi_n$$

Here, an additional term π_{AB} is introduced.

Step 4

$$\frac{T_9}{T_a} = \frac{T_{09}/T_a}{T_{09}/T_9}$$
$$= \frac{T_{09}/T_a}{(p_{09}/p_9)^{(\gamma_g-1)/\gamma_g}}$$

Step 5

From enthalpy balance in main burner, 03 – 04.

$$f = \frac{\tau_\lambda - \tau_r \tau_c}{[\eta_b h_{PR}/(C_{pa}.T_a) - \tau_\lambda]}$$

Now, enthalpy balance in afterburner,

$$(\dot{m}_a + \dot{m}_f)C_{pg}.T_{06} + \eta_{AB}.h_{PR}.\dot{m}_{f\,AB} = (\dot{m}_a + \dot{m}_f + \dot{m}_{f\,AB}).C_{pg}.T_{07}$$

(It may be noted that $T_{05} = T_{06}$ and $T_{07} = T_{09}$)

$$\eta_{AB}.h_{PR}.\dot{m}_{f\,AB} - \dot{m}_{f\,AB}.C_{pg}.T_{07} = (\dot{m}_a + \dot{m}_f)\,[C_{pg}T_{07} - C_{pg}T_{06}]$$

divide by $\dot{m}_a$,

$$\eta_{AB}.h_{PR}.\frac{\dot{m}_{f\,AB}}{\dot{m}_a} - \frac{\dot{m}_{f\,AB}}{\dot{m}_a}C_{pg}.T_{09} = \left(1 + \frac{\dot{m}_f}{\dot{m}_a}\right)[C_{pg}T_{09} - C_{pg}T_{05}]$$

$$\eta_{AB}.h_{PR}.f_{AB} - f_{AB}.C_{pg}.T_{09} = (1+f)(C_{pg}\,T_{09} - C_{pg}.T_{05})$$

$$f_{AB}(h_{PR}\,\eta_{AB} - C_{pg}\,T_{09}) = (1+f)\,(C_{pg}\,T_{09} - C_{pg}\,T_{05})$$

divide by $C_{pa}\,T_a$,

$$f_{AB}\left(\frac{h_{PR}\,\eta_{AB}}{C_{pa}T_a} - \frac{C_{pg}T_{09}}{C_{pa}T_a}\right) = (1+f)\left(\frac{C_{pg}T_{09}}{C_{pa}T_a} - \frac{C_{pg}T_{05}}{C_{pa}T_a}\right)$$

$$f_{AB}\left(\frac{h_{PR}\,\eta_{AB}}{C_{pa}T_a} - \tau_{\lambda\,AB}\right) = (1+f)(\tau_{\lambda\,AB} - \tau_\lambda.\tau_t)$$

as,

$$\tau_\lambda.\tau_t = \frac{C_{pg}.T_{04}}{C_{pa}.T_a}.\frac{C_{pg}T_{05}}{C_{pg}T_{04}}$$

$$= \frac{C_{pg}T_{05}}{C_{pa}T_a}$$

$$f_{AB} = (1+f)\frac{\tau_{\lambda\,AB} - \tau_\lambda\tau_t}{(\eta_{AB}\,h_{PR}/C_{pa}T_a) - \tau_{\lambda\,AB}}$$

Step 6: The power balance between the turbine and compressor is the same as for turbojet engine.

Step 7: Specific thrust can now be evaluated.

Step 8: Thrust specific fuel consumption.

$$S = \frac{(f + f_{AB})}{(F/\dot{m}_a)}$$

Step 9: In propulsive efficiency and thermal efficiency evaluation, along with f, f_{AB} will also be added.

$$\eta_p = \frac{2V_a(F/\dot{m}_a)}{a_a^2[(1+f+f_{AB})((V_9/a_a)^2 - M_a^2)]}$$

$$\eta_T = \frac{a_a^2[(1+f+f_{AB})((V_9/a_a)^2 - M_a^2)]}{2(f+f_{AB})\,h_{PR}}$$

10.4 TURBOFAN-SEPARATE EXHAUST STREAMS

The schematic diagram is given in Figure 9.9 and the T-s diagram for the core/hot stream is given in Figure 10.1. Figure 10.3 gives the T-s diagram for the fan/cold stream taking care of losses. The exit velocity of both the core stream and fan is reduced due to engine losses.

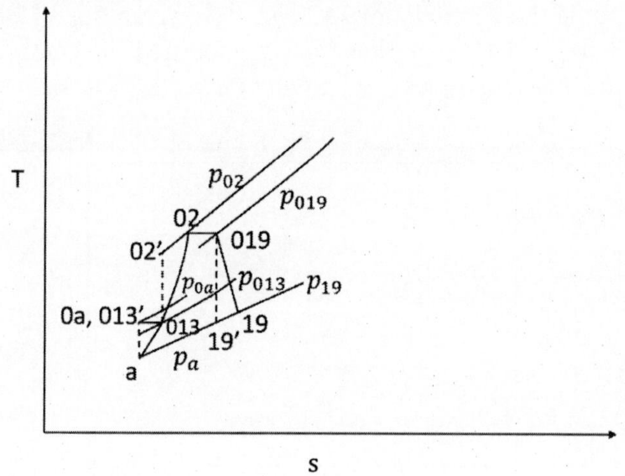

Figure 10.3: T-s Diagram with Losses for the Fan Stream of a Turbofan Engine

Step 1: Uninstalled thrust for fan/cold stream

$$F_F = \dot{m}_F(V_{19} - V_a) + A_{19}(p_{19} - p_a)$$

where, $\dot{m}_F$ is mass flow through the fan stream.

A_{19} is the cross-sectional area of the fan stream nozzle.

V_{19} velocity of the fan jet.

Step 2: Engine core/hot stream

$$\frac{F_h}{\dot{m}_h} = a_a\left[(1+f)\frac{V_9}{a_a} - M_a + (1+f)\frac{R_g}{R_a}\cdot\frac{T_9/T_a}{V_9/a_a}\cdot\left(\frac{1-(p_a/p_9)}{\gamma_a}\right)\right]$$

where, $\dot{m}_h$= mass of air flowing through the core/hot i.e., compressor, combustion chamber and the jet.

$$f = \dot{m}_f / \dot{m}_h$$

Step 3: In power balance between turbine, compressor and fan, the efficiency of all the components and mechanical efficiency, η_m of the coupling should be taken into consideration.

Step 4: Now specific thrust, thrust specific fuel consumption, η_p and η_T can be easily evaluated.

10.5 TURBOFAN WITH AFTERBURNING-MIXED EXHAUST STREAM

The schematic diagram with station numbering for a mixed flow turbofan with afterburner is shown in Figure 9.12 and T-s diagram for an ideal engine is shown in Figure 9.13. This diagram is modified including component efficiencies and pressure losses.

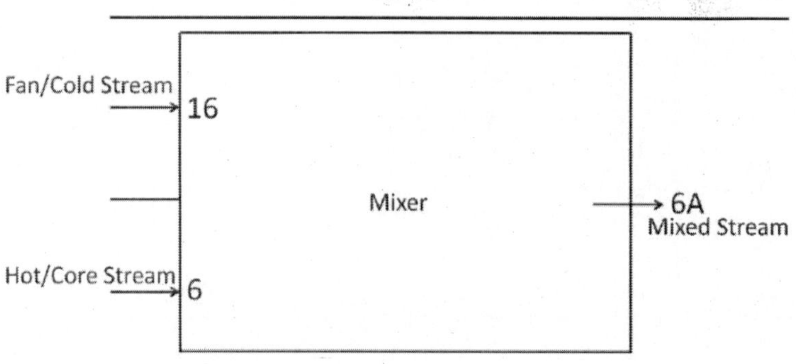

Figure 10.4: Ideal Constant Area Mixer

(*Source:* Mattingly, Jack D., Elements of Gas Turbine Propulsion, McGraw Hill Education, 2017)

The mixed flow turbofan has the following advantages: (i) one variable area exhaust nozzle, (ii) one augmenter (afterburner), and (iii) cold bypass air to cool the afterburner liner. Mixer also improves the performance when the afterburner is off. Figure 10.4 gives the model of an ideal constant area mixture.

$$\dot{m}_6 = \dot{m}_h + \dot{m}_f$$

$$= \text{mass flow rate of air and fuel flowing through the hot stream}$$

$$\dot{m}_{13} = \dot{m}_{16} = \dot{m}_F = \text{mass flow rate of air through fan stream}$$

$6A$ is condition of fluid in the mixed stream

A_6 = Area of the nozzle – hot stream

A_{16} = Area of the nozzle – fan stream

A_{6A} = Area of the mixed flow nozzle

$$A_6 + A_{16} = A_{6A}$$

Apply law of conservation of mass, energy and momentum on mixed flow:

$$\dot{m}_6 + \dot{m}_{16} = \dot{m}_{6A}$$

Temperatures T_{05} and T_{06} are same.

$$\dot{m}_6 C_{p6} T_{06} + \dot{m}_{16} C_{p16} T_{016} = \dot{m}_{6A} C_{p6A} T_{06A}$$

$$p_6 A_6 + \dot{m}_6 V_6 + p_{16} A_{16} + \dot{m}_{16} V_{16} = p_{6A} A_{6A} + \dot{m}_{6A} V_{6A}$$

The properties $C_p, R\ and\ \gamma$ also need to be evaluated for the mix stream.

The engine has single inlet and exhaust. The overall fuel air ratio is equal to the ratio of sum of mass of fuel, $\dot{m}_f$ in main burner and mass of fuel, $\dot{m}_{f\ AB}$ in the afterburner to the total mass of air, $\dot{m}_a$.

$$f_0 = \frac{\dot{m}_f + \dot{m}_{f\ AB}}{\dot{m}_a}$$

also, $\dot{m}_9 = \dot{m}_a + \dot{m}_f + \dot{m}_{f\ AB}$

therefore, $\frac{\dot{m}_9}{\dot{m}_a} = 1 + f_0$

Step 1: Uninstalled thrust

$$\frac{F}{\dot{m}_a} = a_a \left[(1+f_0)\frac{V_9}{a_a} - M_a + (1+f_0)\frac{R_9}{R_a}.\left(\frac{T_9/T_a}{V_9/a_a}\right).\frac{1-(p_a/p_9)}{\gamma_a} \right]$$

Step 2: Thrust specific fuel consumption,

$$S = \frac{f_0}{F/\dot{m}_a}$$

Step 3: The propulsive and thermal efficiencies

The total fuel in the main burner and afterburner is to be accounted.

$$\eta_p = \frac{2V_a(F/\dot{m}_a)}{a_a^2[(1+f_0)(V_9/a_a)^2 - M_a^2]}$$

and

$$\eta_T = \frac{a_a^2[(1+f_0)(V_9/a_a)^2 - M_a^2]}{2f_0.h_{PR}}$$

10.6 TURBOPROP

The advanced turboprop engine technology is developed on two approaches. In one propulsion system program, highly swept propeller blades are connected by an advanced gear box to conventional gas turbine engine. The gear box allows a better match in rotating speeds between low pressure turbine and propeller. In the other approach, the unducted fan (UDF) engine is developed which has two counter rotating rows of highly swept propeller blades that are directly driven by stages of the low-pressure turbine. Lack of gear box reduces engine weight and increases engine reliability.

In real engine analysis, the mechanical efficiency of transmission in high pressure turbine, η_{mH} is introduced. For the low-pressure turbine (free turbine), the mechanical efficiency of transmission, η_{mL} and efficiency factor, η_g in $\dot{W}_{prop}$ is introduced. Thus,

$$\eta_{mL}.\dot{m}_{4.5}.C_{pg}.(T_{04.5} - T_{05}) = \frac{\dot{W}_{\text{prop}}}{\eta_g}$$

The thermal efficiency of a turboprop engine cycle is defined as a ratio of total power produced by the engine to the energy contained in the fuel.

$$\eta_T = \frac{\dot{m}_a.C_{pa}.T_a.C_{tot}}{\dot{m}_f.h_{PR}}$$

The propulsive efficiency of a turboprop engine is defined as a ratio of total power interaction with the vehicle producing propulsive power to the total energy available for producing propulsive power.

$$\eta_P = \frac{2.(\dot{m}_a.C_{pa}.T_a.C_{tot})}{\dot{W}_{\text{prop}}(\dot{m}_9.V_9^2 - \dot{m}_a.V_a^2)}$$

SUMMARY

In real cycle analysis, the component efficiency and losses are taken into consideration. Both the compression and expansion processes are reversible and adiabatic involving increase in entropy. Due to friction, there will be pressure loss in combustion chamber. The working fluid is air in compressor and hot products of combustion in turbine and nozzle. Transmission losses between turbine and compressor also need consideration.

The engines studied include turbojet, turbojet with afterburner, turbofan, turbofan with afterburner and turboprop. The propulsive efficiency, η_P and thermal efficiency, η_T have also been evaluated.

QUESTIONS

1. With neat sketch and T-s diagram, explain the working of a real turbojet engine. Derive the expression for the thrust developed.
2. Draw and describe all thermodynamic processes on the T-s diagram of a real turbojet engine with afterburner.
3. Explain and draw the T-s diagram of a real turbofan with afterburner.
4. How are turbofan engines most efficient for high subsonic aircraft?

CHAPTER 11

Axial Flow Compressor

11.1 INTRODUCTION

A rotating blade row or a rotor changes the total enthalpy of the fluid passing through it either by adding energy to the fluid or adding work to the rotor by taking out energy of the fluid. Enthalpy changes are associated with the pressure changes occurring simultaneously in the fluid. Germans started working on development of axial flow compressors while the British were working on development of centrifugal compressors. For aircraft application, the axial flow compressor was more suitable, as it was able to handle large amounts of air with higher isentropic efficiency. Pressure ratio up to 4:1 was easily obtained from a single stage compressor made of aluminium alloys. Development of titanium alloys resulted in a pressure ratio greater than 8:1 in a single stage. Centrifugal compressor has the advantage that it occupies smaller length than the equivalent axial flow compressor. The axial flow compressor, for jet engines, has a major advantage of handling much larger airflow rate for a given frontal area. Over the years, axial flow compressors have been developed to give pressure ratios exceeding 40:1.

When a fluid passes through a rotating machine both energy transfer and energy transformation take place. Energy transfer means transfer of energy from rotor to fluid or vice versa and energy transformation means change of one form of energy into another form, for example change of kinetic energy to pressure energy in a compressor. In compressors, the energy is transferred from rotor to the fluid while in turbines it is from fluid to the rotor.

The axial flow compressor dominates for large power and centrifugal compressor is restricted to low power requirement as when the flow is too small it is difficult to be handled efficiently by axial blading.

In compressor, the fluid is subjected to adverse pressure gradient and higher the pressure ratio, more difficult to design the compressor. Though the absolute velocity of fluid increases in the rotor blade, the fluid velocity relative to the rotor decreases i.e., there is diffusion within the rotor passages. Therefore, the cross-sectional area is changed as the number of stages progress.

11.2 EULER'S ENERGY EQUATION

Consider the flow of fluid in a rotor as shown in figure 11.1. The fluid particle moves from position 1 at a radius r_1 to position 2 at a radius r_2 ($r_2 > r_1$). A-A is the axis of rotation, and the rotor rotates at a steady angular velocity of ω radians/second.

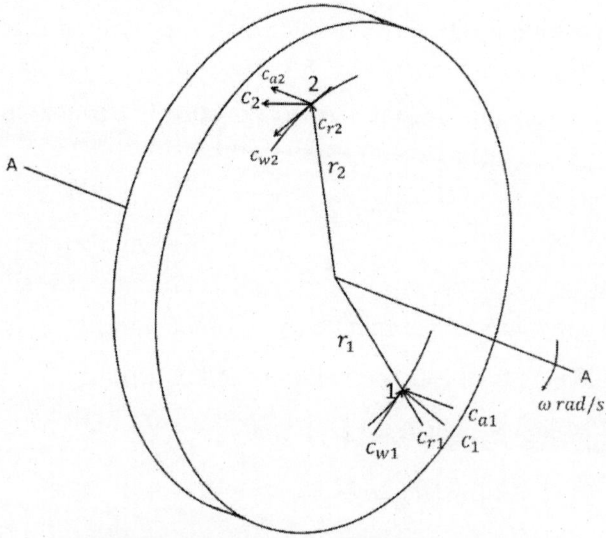

Figure 11.1: Fluid Flow Through a Rotor
(*Source:* Ganesan V., Gas Turbines, Tata McGraw Hill Publishing Company Limited, New Delhi, 2007)

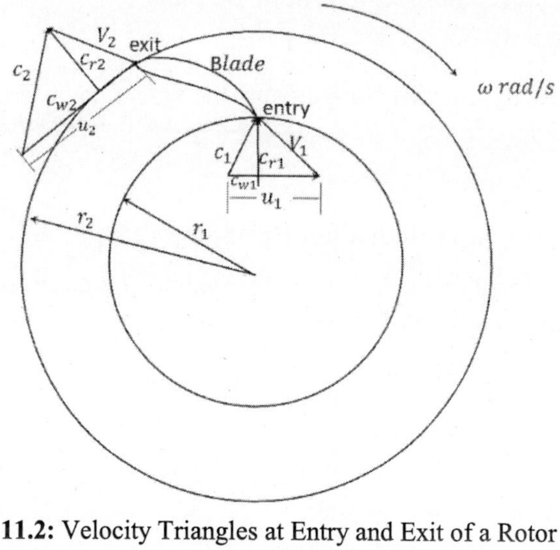

Figure 11.2: Velocity Triangles at Entry and Exit of a Rotor Blade

Let the absolute fluid velocity at position 1 be c_1 and at position 2 be c_2. The velocity c_1 can be divided into three components,

1. c_{a1} : axial component parallel to the axis of rotation
2. c_{r1} : radial component normal to the axis of rotation
3. $c_{\omega 1}$: tangential, also called whirl component normal to radius.

Similarly, at position 2, the three components of c_2 will be c_{a2}, c_{r2} and $c_{\omega 2}$ respectively.

In Figure 11.2, a rotating blade is shown and velocity triangles at entry 1 and exit 2 are drawn.

The angular speed,

$$\omega = \frac{2\pi N}{60}$$

where, N is the revolutions per minute (rpm) of the rotor.

The peripheral velocity at entry and exit, u_1 and u_2 will be

$$u_1 = \frac{\pi N d_1}{60} = \omega . r_1 \quad (d_1 = 2r_1)$$

$$u_2 = \frac{\pi N d_2}{60} = \omega . r_2 \quad (d_2 = 2r_2)$$

where, d_1 and d_2 are diameters of the rotor at position 1 and 2 respectively.

c_1 and c_2 are the absolute velocities, those which will be observed by a person standing outside the rotor. Whereas, the relative velocity vector, V are the ones which will be observed by a person sitting on the rotor.

$$\bar{c} = \bar{u} + \bar{V}$$

The absolute velocity has two components, the tangential or whirl component c_ω and radial component c_r.

Now,

Torque = Rate of change in moment of momentum

It is tangential component of velocity which produces rotation effect. Consider the unit mass flow of fluid.

Angular moment of momentum at inlet = $c_{w1}.r_1$

Angular moment of momentum at exit = $c_{w2}.r_2$

As per Newton's second law of motion, the rate of change of angular momentum is equal to torque produced.

Torque,

$$T = c_{w1}.r_1 - c_{w2}.r_2$$

Therefore, energy transfer, *E* for unit mass of flow per unit time will be,

$$E = T.\omega$$
$$= \omega(c_{w1}r_1 - c_{w2}r_2)$$
$$= c_{w1}u_1 - c_{w2}u_2$$

This is called Euler's energy equation.

When $c_{w1}.u_1$ is more than $c_{w2}.u_2$, the energy transfer is positive and is transferred from fluid to rotor and therefore it is a turbine and when $c_{w2}.u_2$ is more than $c_{w1}.u_1$ then the energy is transferred from rotor to fluid and it is a compressor.

However, the Euler's equation is valid with the following conditions:

1. The flow is steady.
2. The velocity is uniform in every infinitesimal streamline.
3. There should be no discontinuity of pressure.

Now, analyze the velocity diagram, Figure 11.2, at position 2.

$$c_{r2}^2 = c_2^2 - c_{w2}^2$$

also, from other side of the triangle,

$$c_{r2}^2 = V_2^2 - (u_2 - c_{w2})^2$$

eliminate c_{r2}^2,

$$c_2^2 - c_{w2}^2 = V_2^2 - (u_2^2 + c_{w2}^2 - 2c_{w2}.u_2)$$

$$c_{w2}.u_2 = \frac{1}{2}(c_2^2 + u_2^2 - V_2^2)$$

similarly, for inlet, position 1,

$$c_{w1}u_1 = \frac{1}{2}(c_1^2 + u_1^2 - V_1^2)$$

The energy equation can now be written as,

$$E = c_{w1}u_1 - c_{w2}u_2$$
$$= \frac{1}{2}[(c_1^2 - c_2^2) + (u_1^2 - u_2^2) + (V_2^2 - V_1^2)]$$

The term $\frac{1}{2}(c_1^2 - c_2^2)$ refers to change of absolute kinetic energy of fluid and is called impulse effect.

The term $\frac{u_1^2-u_2^2}{2}$ is energy transfer due to centrifugal effect and the third term $\frac{1}{2}(V_2^2 - V_1^2)$ is the energy transfer due to relative kinetic energy of the fluid. This is a reaction effect and gives rise to change in static pressure.

The machines may be classified as axial flow, radial flow or mixed flow:

1. In axial flow machines, the fluid flows parallel to the axis of rotation and therefore, $u_1 \approx u_2$.
2. In centrifugal machines, the flow is radial and there is change in radius at entry and exit of flow.
3. Sometimes the machine may be mixed flow, as none of the axial or radial effect is dominating.

For an axial flow compressor,

$$u_1 = u_2 = U$$

and the Euler's equation will be

$$E = U(c_{w2} - c_{w1})$$

as c_{w2} is more than c_{w1} .

The degree of reaction is a measure of the extent to which the rotor contributes to the overall static pressure rise in the stage and is normally defined as the ratio of static enthalpy rise in the rotor to static enthalpy rise in the stage. This can also be written as:

$$\text{Degree of reaction} = \frac{\text{static temperature rise in the rotor}}{\text{static temperature rise in stage}}$$

A stage consists of a rotor and a stator.

Based on degree of reaction, the machines may be classified as:

1. Impulse machines, where there is no change of static pressure in the rotor and the degree of reaction is zero. An example is Pelton turbine where all pressure energy is converted to kinetic energy before entry of fluid to the turbine. A jet of fluid strikes the blade.
2. Reaction machines are those machines where all energy transfer takes place by change of static pressure in the rotor. The degree of reaction is unity, and an example is lawn sprinkler.

11.3 THEORY

The axial flow compressor consists of a series of stages, each stage consists of a row of rotor blades followed by a row of stator blades. The fluid is initially accelerated by the rotor blades, and then decelerated in the stator blades converting kinetic energy to static pressure. The process is repeated in a number of stages to achieve the desired overall pressure ratio. A notable feature of axial flow compressor is that axial velocity remains approximately constant throughout the compressor. The blades use airfoil section to obtain higher efficiency. The working fluid in an axial flow compressor is normally air. A sketch of a typical stage is given in Figure 11.3.

Figures 11.4 (a), (b) and (c) respectively give a view of the rotor, stator blades fixed with the casing and assembly. One stage of rotor and stator is shown in Figure 11.5 along with the velocity triangles at entry and exit of the rotor blade. The fluid strikes the blade at a velocity c_1 at an angle α_1 and is deflected to an increased velocity c_2 at angle α_2. It may be noted that $\alpha_2 > \alpha_1$. The axial component of velocity c_1 is c_{a1}, and of c_2 is c_{a2} and $c_{a1} = c_{a2} = c_a$. The peripheral velocity U of the rotor at a mid point of blade height remains unchanged. The whirl or tangential components c_{w1} and c_{w2} are important and responsible for transferring energy from rotor to fluid. Figure 11.6 gives the streamlines of the fluid passing through the rotor blade. β_1 and β_2 are rotor blade angles.

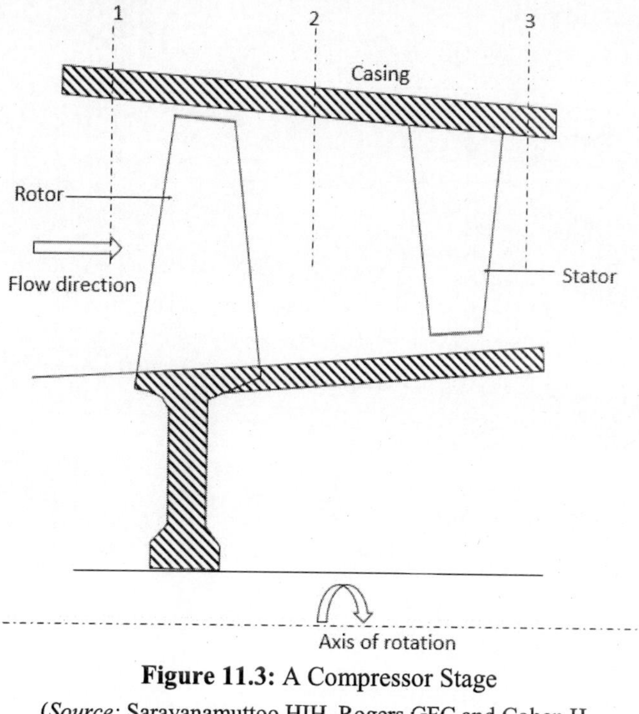

Figure 11.3: A Compressor Stage

(*Source:* Saravanamuttoo HIH, Rogers GFC and Cohen H., Gas Turbine Theory, Pearson Education. 2005)

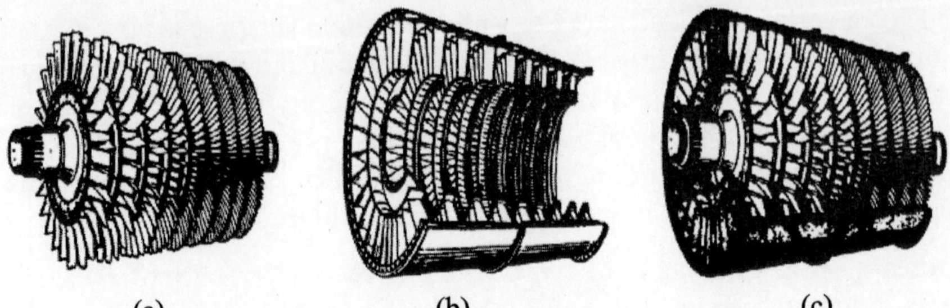

Figure 11.4: (a) Compressor Rotor, (b) Stator, and (c) Assembly

(*Source:* Treager, I.E., Aircraft Gas Turbine Engine Technology, McGraw Hill, New York, Second Edition, 1979)

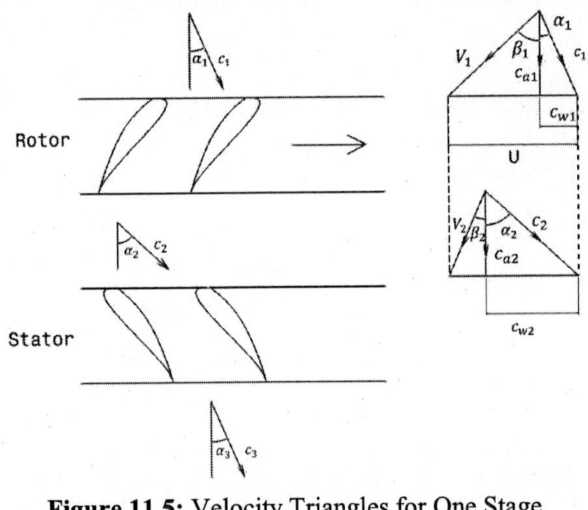

Figure 11.5: Velocity Triangles for One Stage

(*Source:* Saravanamuttoo HIH, Rogers GFC and Cohen H., Gas Turbine Theory, Pearson Education. 2005)

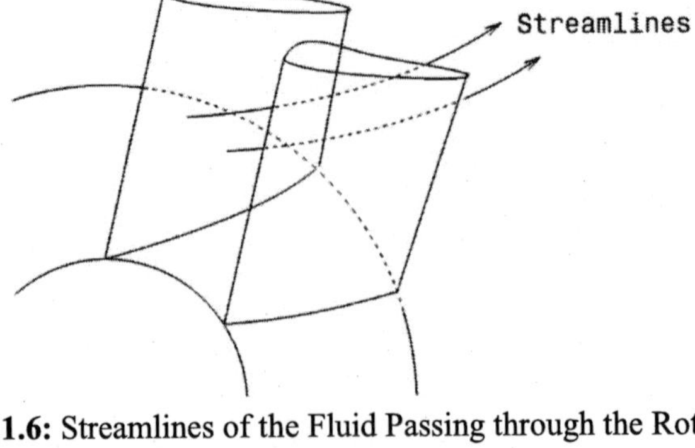

Figure 11.6: Streamlines of the Fluid Passing through the Rotor Blades

The T-s diagram for a stage consisting of rotor and stator is given in Figure 11.7. Position 1–2 is rotor and 2–3 is stator. p_1 and p_{01} are static and total pressures of the inlet fluid to the rotor. The initial velocity of fluid entering the rotor is c_1. In rotor, the static pressure rises to p_2 and fluid velocity to c_2. The total pressure rise in rotor is from 01 to 02. Assuming the process to be adiabatic the enthalpy at 02 and 03, the stator remains the same. In stator, the velocity of the fluid decreases and the static pressure of the fluid increases to p_3. The exit velocity of stator fluid is c_3 at angle α_3. Attempt is made to see that $c_1 = c_3$ and $\alpha_1 = \alpha_3$ so that the fluid is ready for entry to the next stage and symmetrical blading is used.

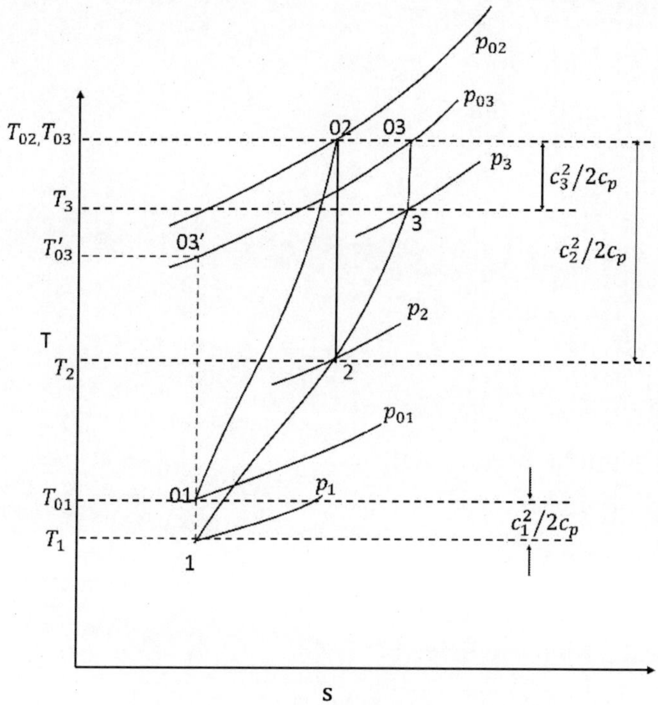

Figure 11.7: T-s Diagram for One Stage

(*Source:* Saravanamuttoo HIH, Rogers GFC and Cohen H., Gas Turbine Theory, Pearson Education. 2005)

The power input to the rotor,

$$W = mC_p(T_{02} - T_{01})$$

There will be some friction loss in the rotor and the stagnation pressure at stator will be p_{03} ($p_{03} < p_{02}$). In stator velocity of the fluid is reduced. Assume,

$$c_{a1} = c_{a2} = c_a$$

From velocity triangles at inlet and exit of the rotor

$$U = c_a \tan\alpha_1 + c_a \tan\beta_1$$

$$\frac{U}{c_a} = \tan\alpha_1 + \tan\beta_1$$

similarly,

$$\frac{U}{c_a} = \tan\alpha_2 + \tan\beta_2$$

from Euler's equation,

$$W = m(Uc_{w2} - Uc_{w1})$$

$$= mU(c_{w2} - c_{w1})$$

where, m is the mass of fluid.

also,

$$c_{w1} = c_a \tan\alpha_1$$

$$c_{w2} = c_a \tan\alpha_2$$

therefore,

$$W = mUc_a(\tan\alpha_2 - \tan\alpha_1)$$

also, from $\frac{U}{c_a}$

$$\tan\alpha_1 + \tan\beta_1 = \tan\alpha_2 + \tan\beta_2$$

$$\tan\alpha_2 - \tan\alpha_1 = \tan\beta_1 - \tan\beta_2$$

therefore,

$$W = mUc_a(\tan\beta_1 - \tan\beta_2)$$

This is the energy input to the rotor.

From T-s diagram, the stage temperature rise,

$$(\Delta T)_{0S} = T_{03} - T_{01}$$

$$= T_{02} - T_{01}$$

also,

$$W = m.C_p.(\Delta T)_{0S}$$

therefore,

$$(\Delta T)_{0S} = \frac{W}{mC_p} = \frac{Uc_a}{c_p}(\tan\beta_1 - \tan\beta_2)$$

The stage isentropic efficiency of compressor,

$$\eta_S = \frac{T'_{03} - T_{01}}{T_{03} - T_{01}}$$

The stage pressure rise,

$$\frac{p_{03}}{p_{01}} = \left(\frac{T'_{03}}{T_{01}}\right)^{\gamma/(\gamma-1)}$$

$$= \left(\frac{T_{01} + \eta_S(T_{03} - T_{01})}{T_{01}}\right)^{\gamma/(\gamma-1)}$$

$$= \left(1 + \frac{\eta_S.(\Delta T_{0S})}{T_{01}}\right)^{\gamma/(\gamma-1)}$$

To minimize the number of stages and obtain a high overall pressure ratio, high temperature rise in stage is required. This can be achieved by,

1. high blade speed
2. high axial velocity
3. high fluid deflection $(\beta_1 - \beta_2)$ in the rotor blades.

11.3.1 Tip Speed

The maximum centrifugal tensile stress, $(\sigma_{ct})_{\max}$ at blade root is given by

$$(\sigma_{ct})_{\max} = \frac{\rho_b.\omega^2}{a}\int_{r_r}^{r_t} a.r\,dr$$

where. ρ_b = density of blade material, r is the radius of the blade from axis of rotation and dr is the elemental thickness. It is assumed that the cross-sectional area, a of the blade is constant from root to tip. The integration is from root, r_r to tip, r_t.

$$\omega = 2\pi N/60$$

where, N is the revolutions per minute.

$$(\sigma_{ct})_{max} = \frac{\rho_b}{2}.(2\pi N/60)^2(r_t^2 - r_r^2)$$

where, r_t and r_r are the radius of the blade at tip and at the root respectively. The tip speed, U_t is given by $2\pi N r_t/60$.

then,

$$(\sigma_{ct})_{max} = \frac{\rho_b}{2}U_t^2\left[1 - \left(\frac{r_r}{r_t}\right)^2\right]$$

The ratio, r_r/r_t is hub to tip ratio. This shows that the centrifugal stress is proportional to square of the tip speed and a reduction of hub to tip ratio increases the blade stress.

The first stage blades, being the longest, are highly stressed, but the later stages blades will be moderately stressed. For tip speeds of about 350 m/s, the stress problems are not critical. Tip speeds of 450 m/s are common in the fans of high bypass ratio turbofans.

11.3.2 High Fluid Deflection in Rotor Blades

The axial velocity should be high, and it helps in providing high flow rate per unit frontal area. Axial velocities up to 200 m/s are used in aero engines. Figure 11.8 gives the velocity diagrams at inlet and exit of a rotor blade on a common base. In axial flow compressor the peripheral velocity of the blade, U and axial component of velocity c_a remains constant.

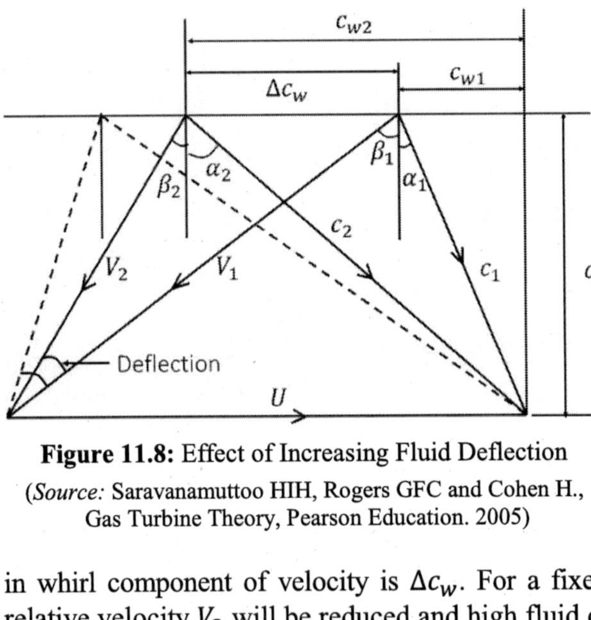

Figure 11.8: Effect of Increasing Fluid Deflection
(*Source:* Saravanamuttoo HIH, Rogers GFC and Cohen H., Gas Turbine Theory, Pearson Education. 2005)

The change in whirl component of velocity is Δc_w. For a fixed β_1, if β_2 is reduced, the relative velocity V_2 will be reduced and high fluid deflection will take place which implies high rate of diffusion.

11.3.3 Work Done Factor

When number of stages is increased, the reduction in work capacity takes place and is accounted by use of work done factor, λ which is always less than unity. The actual temperature rise will now be,

$$\Delta T_{0S} = \frac{\lambda}{C_p} U c_a (\tan\beta_1 - \tan\beta_2)$$

The work done factor is introduced to take care of the distortion of the axial velocity distribution as the number of stages increase. The work done factor is a measure of the ratio of the actual work absorbing capacity of the stage to its ideal value as calculated from the equation.

11.3.4 Degree of Reaction

The degree of reaction is a measure of the extent to which the rotor contributes to the overall static pressure rise in the stage. It is defined as,

$$\Lambda = \frac{\text{static enthalpy rise in the rotor}}{\text{static enthalpy rise in the stage}}$$

For analysis, the following assumptions are made,

1. c_a is constant through the stage
2. The absolute velocity in stage, $c_1 = c_3$ and therefore, $\Delta T_S = \Delta T_{0S}$

ΔT_A *and* ΔT_B denote the static temperature rise in rotor and stator respectively.

$$\Delta T_A = T_2 - T_1$$

$$\Delta T_B = T_3 - T_2$$

The work required by compressor per unit mass of air/fluid.

$$\begin{aligned} W &= C_p(\Delta T_A + \Delta T_B) \\ &= C_p \Delta T_S \\ &= U c_a (\tan\beta_1 - \tan\beta_2) \\ &= U c_a (\tan\alpha_2 - \tan\alpha_1) \end{aligned}$$

All the work input in the stage is in rotor, therefore,

$$\begin{aligned} W &= C_p \Delta T_A + \frac{1}{2}(c_2^2 - c_1^2) \\ &= U c_a (\tan\alpha_2 - \tan\alpha_1) \end{aligned}$$

$$C_p . \Delta T_A = U c_a (\tan\alpha_2 - \tan\alpha_1) - \frac{1}{2}(c_2^2 - c_1^2)$$

from velocity triangle,

$$c_2 = c_a \sec\alpha_2$$

$$c_1 = c_a \sec\alpha_1$$

therefore,

$$C_p\,\Delta T_A = UC_a(\tan\alpha_2 - \tan\alpha_1) - \frac{c_a^2}{2}(\sec^2\alpha_2 - \sec^2\alpha_1)$$

$$= Uc_a\left(\tan\alpha_2 - \tan\alpha_1 - \frac{1}{2}c_a^2(\tan^2\alpha_2 - \tan^2\alpha_1)\right)$$

Now, by definition

$$\Lambda = \frac{\Delta T_A}{\Delta T_A + \Delta T_B}$$

$$= \frac{Uc_a(\tan\alpha_2 - \tan\alpha_1) - \frac{1}{2}c_a^2(\tan^2\alpha_2 - \tan^2\alpha_1)}{Uc_a(\tan\alpha_2 - \tan\alpha_1)}$$

$$= 1 - \frac{c_a}{2U}(\tan\alpha_2 + \tan\alpha_1)$$

also, we know that

$$\frac{U}{c_a} = \tan\alpha_1 + \tan\beta_1$$

$$\frac{U}{c_a} = \tan\alpha_2 + \tan\beta_2$$

by adding the above two,

$$\frac{2U}{c_a} = \tan\alpha_1 + \tan\beta_1 + \tan\alpha_2 + \tan\beta_2$$

$$\tan\alpha_2 + \tan\alpha_1 = \frac{2U}{c_a} - (\tan\beta_1 + \tan\beta_2)$$

therefore,

$$\Lambda = 1 - \frac{c_a}{2U}\left[\frac{2U}{c_a} - (\tan\beta_1 + \tan\beta_2)\right]$$

$$= \frac{c_a}{2U}(\tan\beta_1 + \tan\beta_2)$$

For 50% degree of reaction, put $\Lambda = 1/2$

$$\frac{1}{2} = \frac{c_a}{2U}(\tan\beta_1 + \tan\beta_2)$$

$$\frac{U}{c_a} = \tan\beta_1 + \tan\beta_2$$

also,

$$\frac{U}{c_a} = \tan\alpha_1 + \tan\beta_1$$

$$= \tan\alpha_2 + \tan\beta_2$$

therefore,

$$\tan\alpha_1 = \tan\beta_2 \qquad \text{i.e. } \alpha_1 = \beta_2$$

and $\tan\alpha_2 = \tan\beta_1$ i.e. $\alpha_2 = \beta_1$

also

$$c_a = c_1 \cos\alpha_1 = c_3 \cos\alpha_3$$

as, $c_1 = c_3$

$$\cos\alpha_1 = \cos\alpha_3$$

hence, $\alpha_1 = \alpha_3$

therefore,

$$\alpha_1 = \beta_2 = \alpha_3$$

and $\beta_1 = \alpha_2$

The velocity diagram for these conditions will be symmetrical and blading designed is called symmetrical blading.

also,

$$c_1 = V_2$$

and $V_1 = c_2$

in above analysis, the work done factor is assumed as unity.

From second law of thermodynamics

$$Tds = dh - \frac{dp}{\rho}$$

For incompressible isentropic flow, ds = 0

$$0 = dh - \frac{dp}{\rho}$$

or,

$$(h_2 - h_1) = (p_2 - p_1)/\rho$$

This shows that enthalpy and static pressure changes are related.

We have discussed earlier that the stage temperature rise is given by

$$\Delta T_{0S} = \frac{Uc_a}{c_p}(\tan\beta_1 - \tan\beta_2)$$

also,

$$U = c_a \tan\alpha_1 + c_a \tan\beta_1$$

therefore,

$$c_a \tan\beta_1 = U - c_a \tan\alpha_1$$

therefore,

$$\Delta T_{0S} = \frac{U}{C_p}[U - c_a \tan\alpha_1 - c_2 \tan\beta_2]$$

$$= \frac{U}{C_p}[U - c_a(\tan\alpha_1 + \tan\beta_2)]$$

The angle α_1 is the outlet air angle from the preceding stator and β_2 is the rotor outlet air angle. Both can be considered as constant.

Divide the above equation by U^2

$$\frac{\Delta T_{0S}}{U^2} = \frac{U}{C_p U^2}[U - c_a(\tan\alpha_1 + \tan\beta_2)]$$

$$\frac{C_p(\Delta T_{0S})}{U^2} = 1 - \frac{c_a}{U}(\tan\alpha_1 + \tan\beta_2)$$

Now we define two coefficients,

Flow coefficient,

$$\varphi = C_a/U$$

Stage blade loading coefficient or temperature coefficient,

$$\psi = \frac{C_p(\Delta T_{0S})}{U^2}$$

and,

$$K = \tan\alpha_1 + \tan\beta_2$$

then,

$$\psi = 1 - \varphi K$$

$\psi - \varphi$ plots are drawn and they are useful in designing a compressor.

Cascade is an endless repeating array of airfoils, shown in Figure 11.9. This can be obtained by unwrapping the stationary stator and rotating rotor airfoils. Each cascade passage works as a small diffuser. In good design large static pressure rise is achieved with acceptable pressure losses and flow instabilities due to shock waves or boundary layer separation.

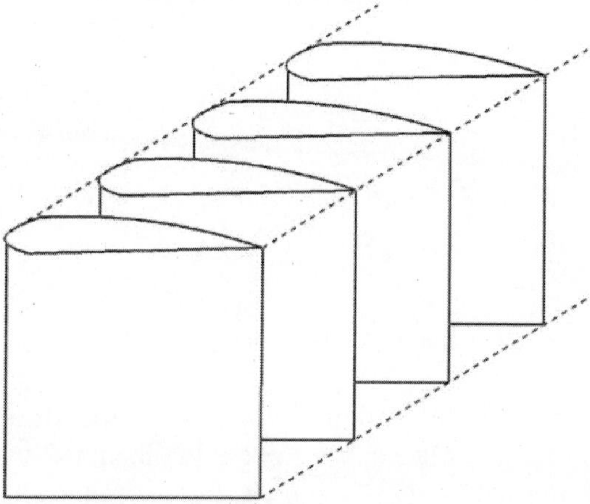

Figure 11.9: Rectilinear Cascade

11.3.5 Diffusion Factor

The total pressure loss of a cascade depends upon the pressure and velocity distribution in airfoil. The upper i.e., suction side of the blade has a large static pressure rise due to deceleration (diffusion) of velocity which can cause boundary layer separation. The amount of diffusion is measured by the diffusion factor on suction side and is defined as the ratio of the difference between the maximum and exit velocity to the average velocity across the airfoil. For axial flow compressor this should be below or equal to 0.6. It may be noted that the diffusion factor for the rotor and stator are different.

The stage exit Mach number M_3 can be estimated from the following equation, $\frac{M_3}{M_1} = \sqrt{\frac{T_1}{T_3}}$ where M_1 is the inlet Mach number of the fluid to the compressor.

There are two unstable flow problems associated with axial flow compressor.

1. Separation of flow from the blade surfaces called stalling, and
2. Complete breakdown of steady flow called surging.

Surging affects the whole machine while stalling is a local phenomenon. Stalling is the separation of flow from the blade surface. Surging of the compressor leads to vibration of the entire machine which can ultimately lead to mechanical failure. To avoid surging, the flow at inlet should be as uniform as possible. The surge can result in large (violent) oscillation on pressure side of the blade.

In Figure 11.10, the rotating stall is shown.

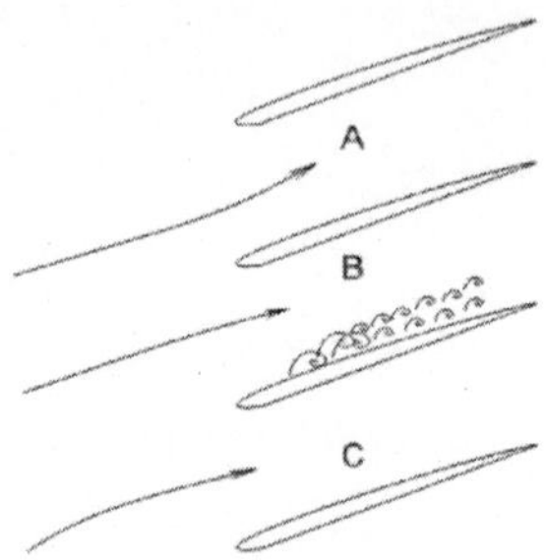

Figure 11.10: Rotating Stall

In this problem, there will be non-uniformity in the flow or geometry of channels between vanes or blades, breakdown in the one channel, say B, causes the air in such a way that channel C receives fluid at reduced angle of incidence and channel A receives at increased angle. Channel A then stalls. This will have the tendency to move the rotor in the opposite direction resulting in vibrations and failure of some parts. The boundary layer separation causes reduction of flow in the channel. As a single blade results in flow diversion, then among the adjacent blades, one is overloaded, and the other is under loaded. This is called rotating stall. The large stresses developed may result in fatigue failure and may be responsible for complete destruction of an entire blade row.

11.4 DESIGN PROCESS

The design process requires the knowledge of typical design values. The following steps may be followed:

1. Selection of rotational speed and annulus dimensions
2. Selection of the number of stages.
3. Calculation of airflow angles for each stage at the mean radius.
4. Calculation of the airflow angle variations from the hub to tip for each stage.
5. Selection of blading.
6. Compressor efficiency estimation.
7. Off-design performance.
8. Rig testing.

The range of various design parameters are given in Table 11.1.

Table 11.1: Axial Flow Compressor Design Parameters

Parameter	***Design Range***
Fan or low-pressure compressor	
Pressure ratio for one stage	1.5–2.0
Pressure ratio for two stages	2.0–3.5
Pressure ratio for three stages	3.5–4.5
Inlet corrected mass flow rate	195–205 $kg/(s.m^2)$
Tip speed	427–457 m/s
Diffusion factor	0.50–0.55
High pressure compressor	
Inlet corrected mass flow rate	175–185 $kg/(s.m^2)$
Hub to tip ratio at inlet	0.60–0.75
Hub to tip ratio at exit	0.90–0.92
Maximum rim speed at exit	396–457 m/s
Flow coefficient	0.45–0.55
Stage loading coefficient	0.30–0.35
Diffusion factor	0.50–0.55

(*Source:* Mattingly, Jack D., Elements of Gas Turbine Propulsion, McGraw Hill Education, 2017)

The preliminary design of multistage axial flow compressor is based on determining each stage flow properties along the compressor mean line. The annulus area at any section is based on the flow properties, T_0, P_0, M, flow angle at the mean radius and total mass flow rate. The mean radius of flow annulus is defined as the average of the tip radius and hub radius. Constant amount of work should be done on fluid passing through a stage i.e., independent of radius. Less turning of fluid is required as the radius increases. The static pressure must increase with the radius to maintain radial equilibrium of the swirling flow. All airfoil and flow properties must vary with radius.

11.4.1 Repeating-Stage, Repeating-Row, Mean-Line Design

The design of an axial flow compressor is complex with many design choices. To simplify design, consider repeating-stage, whose exit velocity and flow angle are equal to those at inlet (repeating-stage). The analysis is carried out for the behaviour of flow at average radius (halfway between hub radius and tip radius), known as mean radius. The number of blades for rotor and stator for any stage should be below 85. The relative velocity Mach number at tip of first stage should be below 1.05 and Mach number at hub of first stage should be below unity.

11.5 COMPRESSOR CHARACTERISTICS

The performance of a compressor is given by curves of delivery pressure and temperature plotted against mass flow for various fixed values of rotation speed. We know that the fundamental dimensions are mass, length and time. A function of all the variables $D, N, m, p_{01}, p_{02}, RT_{01}$ and RT_{02} can be written and equated to zero.

$$\text{Function } (D, N, m, p_{01}, p_{02}, RT_{01} \text{ and } RT_{02}) = 0$$

where, D is characteristic linear dimension usually impeller diameter and N is rotation speed. Using Pi theorem for dimensional analysis, the following non-dimensional terms will emerge.

$$\frac{p_{02}}{p_{01}}, \frac{T_{02}}{T_{01}}, \frac{m\sqrt{RT_{01}}}{D^2 p_{01}}, \frac{ND}{\sqrt{RT_{01}}}$$

For a particular machine and fluid, the terms D and R can be omitted and the groups can be written as,

$$\text{Function}\left(\frac{p_{02}}{p_{01}}, \frac{T_{02}}{T_{01}}, \frac{m\sqrt{T_{01}}}{p_{01}}, \frac{N}{\sqrt{T_{01}}}\right) = 0$$

where, the quantities $\frac{m\sqrt{T_{01}}}{p_{01}}$ and $\frac{N}{\sqrt{T_{01}}}$ are not actually dimensionless but they are treated as non-dimensional mass flow and rotational speed respectively. The terms total pressure ratio and total temperature ratio (T_{02}/T_{01}) are dimensionless i.e., non-dimensional parameters and m is mass flow of fluid.

The performance characteristics of an axial flow compressor are shown in figure 11.11 (a) and (b). The total or stagnation pressure ratio (p_{02}/p_{01}) and isentropic efficiency, η are plotted against non-dimensional mass flow $m\sqrt{T_{01}}/p_{01}$ for fixed values of non-dimensional speed $N/\sqrt{T_{01}}$. At high rotational speeds the constant speed lines become very steep and ultimately may be vertical. The same limitation occurs at either end of $N/\sqrt{T_{01}}$ due to surging and choking. The surge line in figure 11.11 (a) gives the locus of unstable operation of the compressor at a given rpm. The efficiency varies with mass flow at a given speed in a similar manner to the pressure ratio, but the maximum value is approximately the same at all speeds (Figure 11.11 (b)).

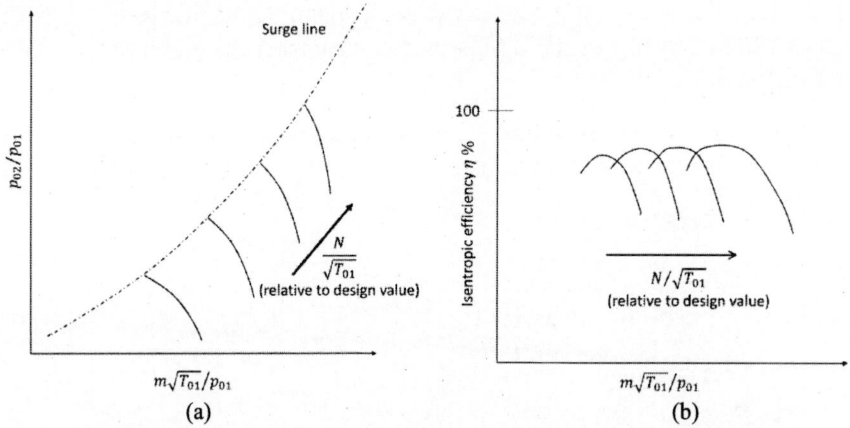

Figure 11.11: (a) and (b) Axial Flow Compressor Characteristics

11.6 CENTRIFUGAL COMPRESSOR

The centrifugal compressor consists of a stationary casing containing a rotating impeller and a diffuser. The impeller increases the velocity of the air and in diffuser the air is decelerated which results in rise of static pressure. The diffuser may be vaneless or with vanes. Figure 11.12 gives two views of a typical centrifugal compressor.

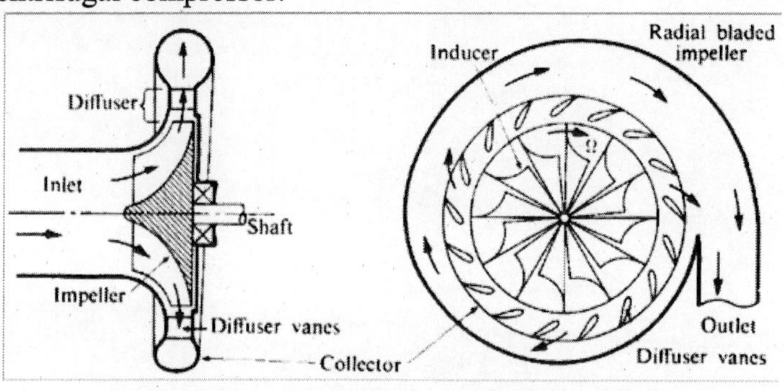

Figure 11.12: A Typical Centrifugal Compressor

(*Source:* Hill P.G. and Peterson C.R., Mechanics and Thermodynamics of Propulsion, Addison-Wesley Publishing Company, 1992)

The major advantages of a centrifugal compressor are:

1. It occupies smaller length,
2. operates more efficiently over a wide range of mass flow rate at any rotational speed, and
3. the stage pressure ratio for centrifugal compressor is higher than the equivalent axial flow compressor.

In impeller, the energy transfer takes place resulting in increase of kinetic energy of fluid and rise in static pressure. A further rise in pressure takes place in the diffuser.

In compressor the energy transfer (E/W) is always from rotor to the fluid and the Euler's equation is given as,

$$W = E = c_{w2}u_2 - c_{w1}u_1$$

also,

$$W = E = \frac{1}{2}\left((c_2^2 - c_1^2) + (u_2^2 - u_1^2) + (V_1^2 - V_2^2)\right)$$

$$W = \text{Work done}$$
$$= E$$
$$= \text{Energy transfer}$$

The velocity diagrams for a centrifugal compressor are given in Figure 11.13.

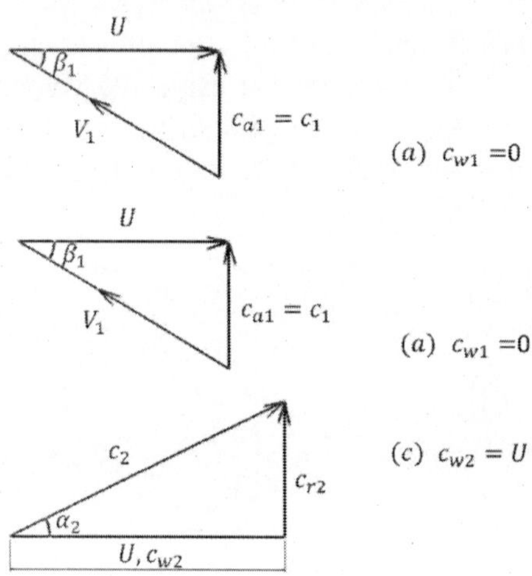

Figure 11.13: The Velocity Diagrams at Inlet and Exit of the Impeller for a Centrifugal Compressor

For an ideal compressor, the air enters the compressor directly from the atmosphere without any tangential/whirl component i.e., $c_{w1} = 0$ (Figure 11.13 (a)) and the air leaves the impeller with a tangential velocity equal to the impeller velocity, $c_{w2} = U$ (Figure 11.13(c)), no slip condition is assumed. Slip factor, σ is defined as c_{w2}/U. Smaller the slip, means c_{w2} more nearly approaches U. Figure 11.13(b) shows the conditions with slip.

therefore, for unit mass flow of air, theoretical torque = $c_{w2} . r_2$

If ω is angular velocity, the work done on air

$$W = c_{w2}.r_2.\omega$$
$$= c_{w2}.U$$

For no slip condition, $c_{w2} = U$

therefore, $W = U^2$

If slip factor σ and power input factor P_{if} are introduced which takes care the friction and other losses, the actual power input to the compressor will be more than the theoretical value.

$$W = \sigma.P_{if}.U^2$$

The power input factor represents the increase in the work input. The normal value for P_{if} is between 1.035 to 1.04.

If T_{01}, T_{02}, T_{03} are the total/stagnation temperatures at inlet of impeller, exit of impeller and exit of diffuser respectively. The work done,

$$W = C_p(T_{03} - T_{01})$$
$$= \sigma P_{if} U^2$$

as $T_{02} = T_{03}$

$$W = C_p(T_{02} - T_{01})$$

also,

$$\frac{p_{03}}{p_{01}} = \left(\frac{T'_{03}}{T_{01}}\right)^{[\gamma/(\gamma-1)]}$$

The compressor efficiency, η_c will be

$$\eta_c = \frac{T'_{03} - T_{01}}{T_{03} - T_{01}}$$

$$(T_{03} - T_{01})\eta_c = T_{01}\left(\frac{T'_{03}}{T_{01}} - 1\right)$$

$$T'_{03} = T_{01}\left[1 + \frac{\eta_c(T_{03} - T_{01})}{T_{01}}\right]$$

therefore,

$$\frac{p_{03}}{p_{01}} = \left[1 + \eta_c.\frac{\sigma P_{if} U^2}{C_p T_{01}}\right]^{[\gamma/(\gamma-1)]}$$

The T-s diagram for a centrifugal compressor is given in Figure 11.14.

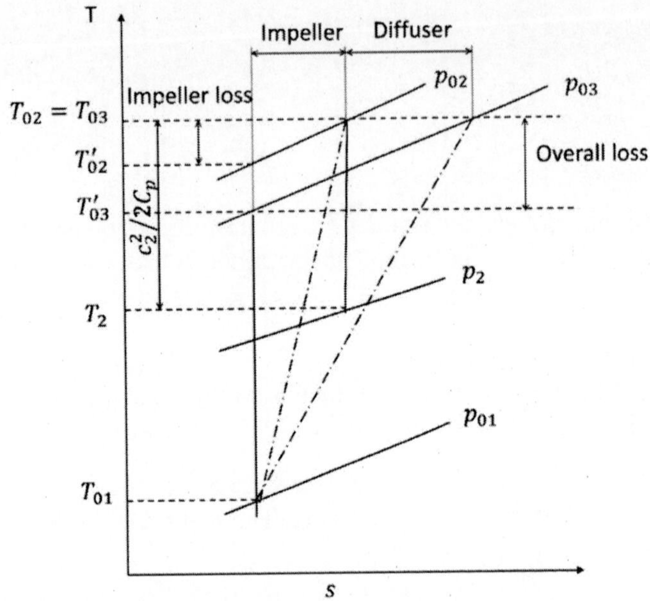

Figure 11.14: T-s Diagram for a Stage-Centrifugal Compressor

(*Source:* Saravanamuttoo HIH, Rogers GFC and Cohen H., Gas Turbine Theory, Pearson Education, 2005)

It may be noted that power input, W is only into the impeller and there is no power requirement in the diffuser.

SUMMARY

A rotating blade or a rotor changes the enthalpy of the fluid passing through it either by adding energy to the fluid or by taking energy of the fluid. Enthalpy changes are associated with pressure changes occurring simultaneously in the fluid. The axial flow compressor for jet engines has a major advantage of handling much larger air flow rate for a given frontal area. In compressors, the energy is transferred from rotor to the fluid while in turbines it is from fluid to the rotor.

The Euler's equation for energy transfer for a turbine for unit mass flow per unit time is:

$$E = c_{w1}u_1 - c_{w2}u_2$$

as, $c_{w1}u_1$ will always be more than $c_{w2}u_2$.

The Euler's equation for a compressor is,

$$E = c_{w2}u_2 - c_{w1}u_1$$

as, $c_{w2}u_2$ will always be more than $c_{w1}u_1$.

For an axial flow compressor, $u_1 = u_2 = U$ and the energy equation can be written as:

$$E = U(c_{w2} - c_{w1})$$

The degree of reaction is a measure of the extent to which the rotor contributes to the overall static pressure rise in the stage.

$$\text{Degree of reaction}=\frac{\text{static temperature rise in the rotor}}{\text{static temperature rise in the stage}}$$

The turbomachines can be classified based on degree of reaction. Impulse machines are those machines where the degree of reaction is zero and an example is Pelton turbine. Reaction machines are those machines where all energy transfer takes place in the rotor and the degree of reaction is unity.

The energy input to the rotor for an axial flow compressor stage is:

$$W = mUc_a(\tan\beta_1 - \tan\beta_2)$$

and the stage pressure rise,

$$\frac{p_{03}}{p_{01}} = \left(1 + \frac{\eta_s.(\Delta T_{0S})}{T_{01}}\right)^{\gamma/(\gamma-1)}$$

When the number of stages is increased, the reduction in work capacity takes place and is accounted using work done factor, λ.

$$(\Delta T_{0S}) = \frac{\lambda}{C_p} Uc_a(\tan\beta_1 - \tan\beta_2)$$

For 50% reaction,

$$\frac{U}{c_a} = \tan\beta_1 + \tan\beta_2$$

and,

$$\alpha_1 = \beta_2$$

$$\beta_1 = \alpha_2$$

It is desirable that after a stage, $c_1 = c_3$ and $\alpha_1 = \alpha_3$ so that symmetrical blading can be designed.

Two coefficients are defined,

Flow coefficient, $\varphi = c_a/U$

Temperature coefficient, $\psi = \frac{C_p\,(\Delta T_{0S})}{U^2}$

There are two unstable flow properties, one is stalling, and the other is surging.

The diffusion factor on suction side of the blade is defined as the ratio of the difference between maximum and exit velocity to the average velocity across the airfoil. The diffusion factor may be different for rotor and stator. A brief description of centrifugal compressor is given.

EXAMPLES

Examples 11.1

Calculate the number of stages required for an axial flow compressor of 50% reaction design. The input data are:

Rotor inlet angle, $\beta_1 = 50°$, rotor outlet angle, $\beta_2 = 15°$, pressure ratio = 5: 1, overall isentropic efficiency = 87%, inlet air temperature = $300\ K$, blade speed = $210\ m/s$, work done factor =1. Assume that the blade speed and axial velocity are constant throughout the stage. ($\gamma = 1.4, C_p = 1.005\ kJ/(kg.K)$)

Solution

Increase in temperature in stage,

$$\Delta T_{0s} = \frac{\lambda\, U\, c_a}{C_p}(\tan\beta_1 - \tan\beta_2)$$

as, $\lambda = 1$

$$\Delta T_{0s} = \frac{U\, c_a}{C_p}(\tan\beta_1 - \tan\beta_2)$$

For 50% reaction design,

$$\frac{U}{c_a} = \tan\beta_1 + \tan\beta_2$$

$$c_a = \frac{210}{\tan 50° + \tan 15°} = \frac{210}{1.1917 + 0.2679}$$

$$= \frac{210}{1.4596} = 143.87\ m/s$$

$$\Delta T_{0s} = \frac{1 \times 210 \times 143.87(\tan 50 - \tan 15)}{1.005 \times 10^3}$$

$$= \frac{1 \times 210 \times 143.87 \times 0.9238}{1005} = 27.77\ K$$

therefore, overall temperature rise in compressor,

$$(\Delta T_0)_{total} = \frac{T_{01}}{\eta_c}\left[\left(\frac{p_{03}}{p_{01}}\right)^{(\gamma-1)/\gamma} - 1\right]$$

$$(\Delta T_0)_{total} = \frac{300}{0.87}\left[5^{\left(\frac{1.4-1}{1.4}\right)} - 1\right]$$

$$= \frac{300}{0.87}(5^{0.286} - 1)$$

$$= \frac{300}{0.87}(0.5845)$$

$$= 201.57$$

Therefore, number of stages

$$= \frac{(\Delta T_0)_{total}}{\Delta T_{0s}} = \frac{201.57}{27.77} = 7.26$$

say, ≈ **8** stages

Examples 11.2

Estimate the blade angles for a 8 stage axial flow compressor which provides an overall pressure ratio of 6:1. Take inlet air temperature as 300 K. The work is equally divided between the stages. A 50% reaction is used with blade speed of 200 m/s and constant axial velocity of 150 m/s. Assume work done factor as unity. The overall isentropic efficiency of compressor is 85%. ($\gamma = 1.4$, $C_p = 1.005\ kJ/kg.K$).

Solution

The overall temperature rise,

$$(\Delta T_0)_{total} = \frac{T_{01}}{\eta_c}\left[\left(\frac{p_{03}}{p_{01}}\right)^{(\gamma-1)/\gamma} - 1\right]$$

$$= \frac{300}{0.85}(6^{0.286} - 1)$$

$$= \frac{300}{0.85} \times 0.6694$$

$$= 236.25\ K$$

Therefore, stage temperature rise

$$(\Delta T_{0s}) = \frac{236.25}{8}\ K = 29.53\ K$$

also,

$$(\Delta T_{0s}) = \frac{\lambda\, U\, c_a}{C_p}(\tan\beta_1 - \tan\beta_2)$$

$$\lambda = 1$$

$$(\tan\beta_1 - \tan\beta_2) = \frac{1.005 \times 10^3 \times 29.53}{200 \times 150}$$

$$= 0.9892$$

For 50% reaction,

$$(\tan\beta_1 + \tan\beta_2) = \frac{U}{c_a} = \frac{200}{150}$$

$$= 1.3333$$

therefore,

$$\tan\beta_1 = \frac{0.9892 + 1.3333}{2} = 1.1612$$

$$\tan\beta_2 = \frac{1.3333 - 0.9892}{2} = 0.1720$$

$$\boldsymbol{\beta_1 = 49.3^\circ}$$

$$\boldsymbol{\beta_2 = 9.8^\circ}$$

QUESTIONS

1. With the help of T-s diagram, describe the functioning of a general axial flow compressor.
2. Draw the velocity triangles for one stage of an axial flow compressor and show that the power input is,
   $W = mUC_a(\tan\beta_1 - \tan\beta_2)$
   where, the symbols have the usual meaning.
3. Explain the following, applied to turbomachines:
   (a) Work done factor
   (b) Flow coefficient
   (c) Temperature drop coefficient
   (e) Degree of reaction
   (f) Loading coefficient
   (g) Symmetrical blades
4. Define the phenomenon of surging and stalling in axial flow compressor.

CHAPTER 12

Axial Flow Turbine

12.1 INTRODUCTION

The turbine is reverse of a compressor. The turbine operates under favourable pressure gradient. This allows greater angular changes, greater pressure changes, greater energy changes and higher efficiency. The stress associated with turbine blades is higher because of higher work and temperatures. The turbine handles products of combustion at very high temperature. The materials of construction of turbine blades are important. In axial flow turbine the stator comes first, usually called nozzles, followed by rotor. The nozzle blades are facing maximum stress because the hot products of combustion hit the nozzles.

In stator or nozzle, the fluid is accelerated, and the static pressure is decreased. The tangential component of velocity increases which produces torque on rotor blades. The absolute velocity of fluid decreases in rotor. A set of a nozzle and rotor is called a stage.

12.2 THEORY

A turbine stage is shown in Figure 12.1. It consists of a nozzle, N (1–2) and rotor, R (2–3).

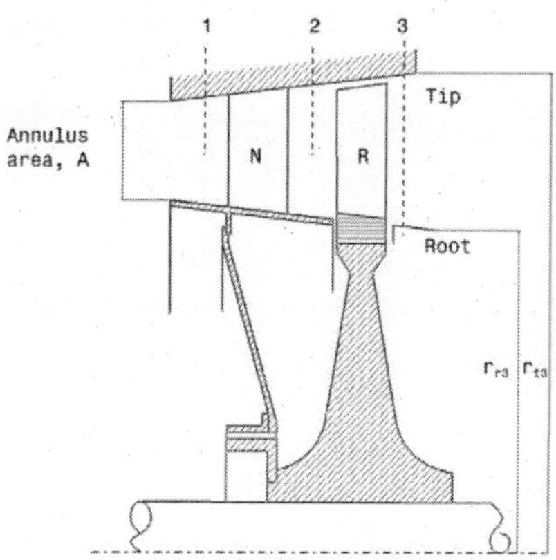

Figure 12.1: Axial Flow Turbine Stage

(*Source:* Saravanamuttoo HIH, Rogers GFC and Cohen H., Gas Turbine Theory, Pearson Education. 2005)

Figure 12.2 shows the velocity diagrams of nozzle inlet and rotor inlet and outlet and Figure 12.3 shows the velocity diagrams of rotor inlet and outlet on a common peripheral velocity, U. Figure 12.4 gives the T-s diagram for a reaction stage.

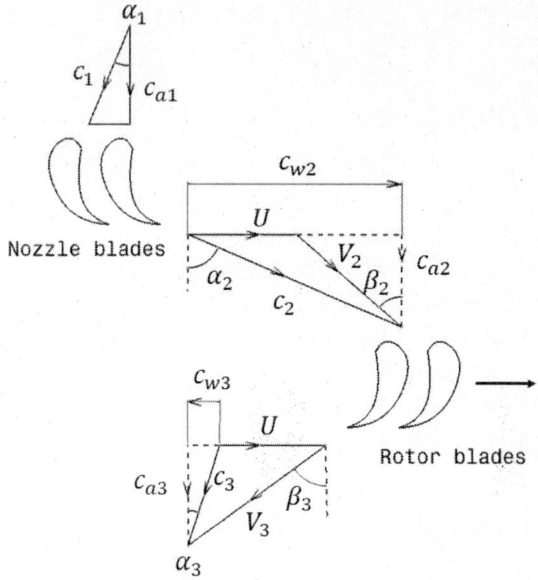

Figure 12.2: Velocity Diagrams at Nozzle and Rotor

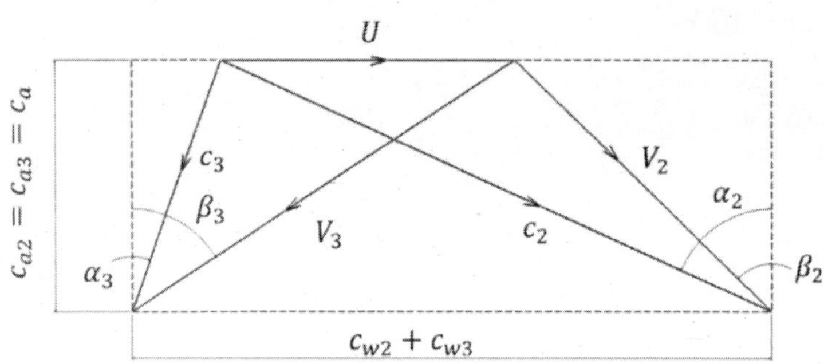

Figure 12.3: Velocity Diagram on a Common Peripheral Velocity, U

In a single stage turbine c_1 will be axial i.e., $\alpha_1 = 0$ and $c_1 = c_{a1}$. The gas enters the nozzle at a static pressure p_1, total pressure p_{01}, temperature T_1 and velocity c_1, is expanded to p_2, p_{02}, T_2 *and* increased velocity c_2. The rotor blade angles are β_2 and β_3. V_2 and V_3 are relative velocities. U is the blade speed. The gas leaves the rotor blade at p_3, p_{03}, T_3 and velocity c_3. The angle α_3 is called swirl angle. The analysis is carried out at mean diameter of the annulus.

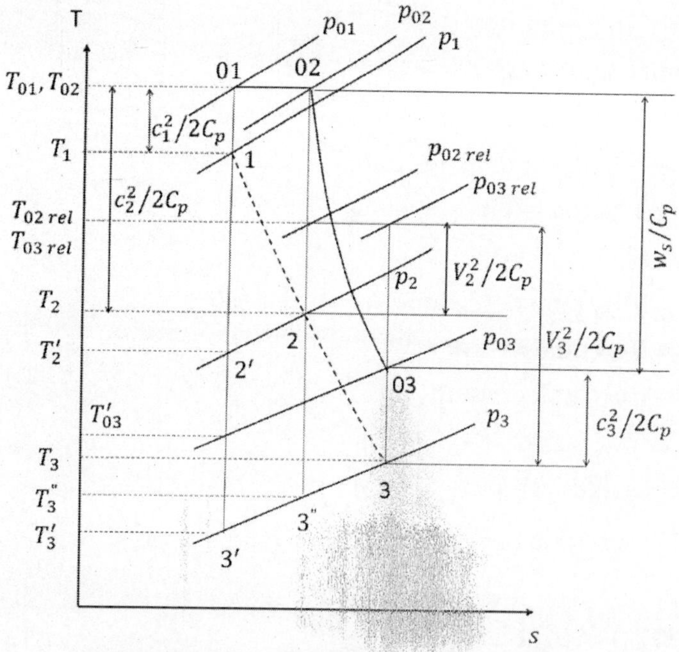

Figure 12.4: T-s Diagram for a Reaction Stage
(*Source:* Saravanamuttoo HIH, Rogers GFC and Cohen H., Gas Turbine Theory, Pearson Education. 2005)

The term $(c_{w2} + c_{w3})$ is the change in whirl or tangential component of velocity which produces the useful torque.

In a multistage turbine c_1 and α_1 will be equal to c_3 and α_3 so that the same blade can be used in subsequent stages. Then it is called repeating stage. It is assumed that the ratio of blade tip radius to root radius is low i.e., the blades are short. For long blades, three dimensional effects need consideration. It is further assumed that the axial velocity is constant through the rotor though it will be changing which will produce some axial thrust on the rotor. Also, the density of the fluid will decrease during expansion through the turbine stage. From figure 12.3, it can be seen that

$$U = c_a \tan\alpha_2 - c_a \tan\beta_2$$

$$= c_a \tan\beta_3 - c_a \tan\alpha_3$$

$$\text{or, } \frac{U}{c_a} = \tan\alpha_2 - \tan\beta_2$$

$$= \tan\beta_3 - \tan\alpha_3$$

The stage work output per unit mass flow, from Euler's equation,

$$W_s = c_{w2}.U - (-c_{w3}U)$$
$$= U(c_{w2} + c_{w3})$$

also,

$$\tan\alpha_2 - \tan\beta_2 = \tan\beta_3 - \tan\alpha_3$$
$$\tan\alpha_2 + \tan\alpha_3 = \tan\beta_2 + \tan\beta_3$$

therefore,

$$W_S = U(c_a \tan\alpha_2 + c_a \tan\alpha_3)$$
$$= Uc_a(\tan\alpha_2 + \tan\alpha_3)$$
$$= Uc_a(\tan\beta_2 + \tan\beta_3)$$

also,

$$W_s = C_p.(\Delta T_{0S})$$

where, ΔT_{0S} is stagnation temperature drop during stage.

therefore,

$$C_p(\Delta T_{0S}) = Uc_a(\tan\beta_2 + \tan\beta_3)$$

When, $c_1 = c_3$, the static temperature change is same as total temperature change.

Some typical values for products of combustion are:

$$C_p = 1.148\frac{kJ}{kg.K}$$

$$\gamma = 1.33$$

$$\frac{\gamma}{(\gamma - 1)} = 4$$

$$R = 0.287\frac{kJ}{kg.K}$$

From T-s diagram, Figure 12.4

$$\frac{T_{01}}{T'_{03}} = \left(\frac{p_{01}}{p_{03}}\right)^{(\gamma-1)/\gamma}$$

Turbine isentropic stage efficiency,

$$\eta_s = \frac{T_{01} - T_{03}}{T_{01} - T'_{03}}$$

therefore, stage total temperature change

$$\begin{aligned}\Delta T_{0S} &= T_{01} - T_{03}\\ &= \eta_s(T_{01} - T'_{03})\\ &= \eta_s T_{01}\left(1 - \frac{T'_{03}}{T_{01}}\right)\\ &= \eta_s T_{01}\left(1 - \frac{1}{(T_{01}/T'_{03})}\right)\\ &= \eta_s . T_{01}\left(1 - \frac{1}{(p_{01}/p_{03})^{(\gamma-1)/\gamma}}\right)\end{aligned}$$

Now, we define three dimensionless numbers which will be useful in turbine design. The first is work capacity of a stage also called blade loading coefficient or temperature drop coefficient, ψ.

$$\begin{aligned}\psi &= \frac{C_p(\Delta T_{0S})}{U^2}\\ &= \frac{C_a}{U}(\tan\beta_2 + \tan\beta_3)\end{aligned}$$

Second, the degree of reaction, Λ

$$\Lambda = \frac{T_2 - T_3}{T_1 - T_3}$$

where c_a is constant and $c_3 = c_1$

$$\begin{aligned}C_p(T_1 - T_3) &= C_p(T_{01} - T_{03})\\ &= Uc_a(\tan\beta_2 + \tan\beta_3)\end{aligned}$$

In terms of relative velocity, V, using steady flow energy equation:

$$\begin{aligned}C_p(T_2 - T_3) &= \frac{1}{2}(V_3^2 - V_2^2)\\ &= \frac{1}{2}c_a^2(sec^2\beta_3 - sec^2\beta_2)\\ &= \frac{1}{2}c_a^2(\tan^2\beta_3 - \tan^2\beta_2)\end{aligned}$$

therefore,

$$\Lambda = \frac{c_a}{2U}(\tan\beta_3 - \tan\beta_2)$$

The third dimensionless parameter is flow coefficient, ϕ, defined as,

$$\phi = \frac{c_a}{U}$$

therefore,

$$\psi = \phi(\tan\beta_2 + \tan\beta_3)$$

and

$$\Lambda = \frac{\phi}{2}(\tan\beta_3 - \tan\beta_2)$$

For 50% degree of reaction, put $\Lambda = 0.5$

$$\frac{1}{2} = \frac{\phi}{2}(\tan\beta_3 - \tan\beta_2)$$

or,

$$\frac{1}{\phi} = \tan\beta_3 - \tan\beta_2$$

Compare this with

$$\frac{U}{c_a} = \tan\alpha_2 - \tan\beta_2 = \frac{1}{\phi}$$

$$= \tan\beta_3 - \tan\alpha_3 = \frac{1}{\phi}$$

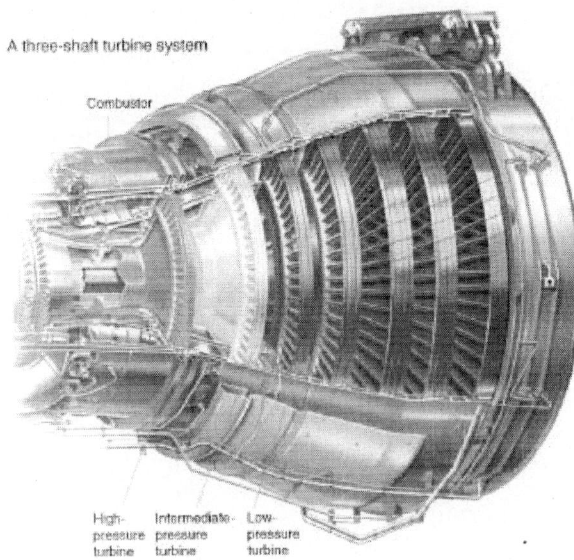

Figure 12.5: A View of an Axial Flow Turbine

(*Source:* Farokhi Saeed, Aircraft Propulsion, Wiley, Second Edition, 2014)

we get, $\beta_3 = \alpha_2$ and $\beta_2 = \alpha_3$ and the velocity diagram becomes symmetrical.

The velocity ratio, VR is defined as the ratio of the rotor speed, U to the velocity equivalent of the change in stage total enthalpy.

$$VR = \frac{U}{\sqrt{2C_p(\Delta T_{0S})}}$$

The velocity ratio is used by some turbine designers in place of stage loading coefficient, ψ. The velocity ratio at the mean radius ranges between 0.5 and 0.6 for modern aircraft gas turbine engines. Figure 12.5 gives a view of an axial flow turbine.

12.3 DESIGN PROCESS

The design process requires knowledge of the range of design parameters. Table 12.1 gives this range for an axial flow turbine. Over the years, there has been increase in turbine inlet temperature, mass flow rate and output power. The stage loading coefficient has not changed appreciably between the designs. Due to increase in turbine inlet temperature, with the use of available materials, extensive use of cooling air has been incorporated. By increasing

Table 12.1: Range of Axial Flow Turbine Design Parameters

Parameter	***Design Range***
High pressure turbine	
Maximum AN^2	$4 - 5 \times 10^{10}$ inch^2.rpm^2
Stage loading coefficient	1.4 – 2.0
Exit Mach number	0.40 – 0.50
Exit swirl angle (deg.)	0–20
Low pressure turbine	
Inlet corrected mass flow rate	195–215 $kg/(s.m^2)$
Hub to tip ratio at inlet	0.35–0.50
Maximum stage loading at hub	2.4
Exit Mach number	0.40–0.50
Exit swirl angle (deg.)	0–20

(*Source:* Mattingly, Jack D., Elements of Gas Turbine Propulsion, McGraw Hill Education, 2017)

stage loading coefficient, the number of stages can be reduced. The design process involves the following steps:

1. Selection of rotational speed and annulus dimensions.
2. Selection of the number of stages.
3. Calculation of airflow angles for each stage at the mean radius.
4. Calculation of the airflow angle variations from the hub to tip for each stage.
5. Selection of blade material.
6. Selection of blading.
7. Selection of turbine cooling, if required.
8. Verification of turbine efficiency.
9. Off-design performance.
10. Rig testing.

The design process is mostly iterative, requiring the return to an earlier step when prior assumptions are found to be invalid. When the required stage loading coefficient, ψ is more than 2.0, multistage turbine design is required. It is assumed that total temperature drop of turbine is evenly distributed in the stages.

$$(\Delta T_0)_{\text{turbine}} = \text{number of stages} \times (\Delta T_{0S})$$

For each stage, same stage loading coefficient, ψ and same degree of reaction, Λ for the same rotor speed, U is assumed. This is valid for 2–3 stages design. The Mach number at entry to the rotor to the first stage may be slightly supersonic but in subsequent stages it is kept subsonic. The stage loading of the first stage is slightly more than the stage loading of the other stages.

The high stresses of the component determine the rotational speed of the shaft. For a low-pressure spool, the first stage normally dictates the rotational speed as this has greatest AN^2, where A is annulus area and N is rotational speed.

In single stage design, the degree of reaction at hub may be negative. For a reasonable stage loading coefficient, ψ, a non-zero exit swirl is assumed. Effort is made to have moderate exit swirl angle, α_3 and positive reaction at subsonic relative velocity Mach number at the tip. The value of AN^2 should be within the acceptable limits. The axial chord and number of blades for the stator or rotor depend on many factors such as flow through the blades, vibrations, blade attachment, etc.

In a two-stage design, the stage loading coefficients, ψ are lower and the temperature ratios are higher than those for single stage design. This results in higher reaction, less turning of the flow and lower loss coefficients. Low supersonic value of Mach number 1.05 for inlet for first stage is selected and relative velocity Mach number below 0.9 is maintained.

For zero reaction axial flow turbine, with constant axial velocity, $T_3 = T_2$ and $\beta_2 = \beta_3$ i.e., there is no enthalpy change in the rotor. The turbine is then called the impulse turbine. For zero exit swirl, $\alpha_3 = 0$ and $c_{w3} = 0$. In aircraft gas turbines, thrust to weight ratio must be balanced. Weight can be reduced by increasing stage loading, thereby reducing the number of stages but this may result in lower stage efficiency.

12.3.1 Blade Spacing

The tangential force on the blades results in pressure side and suction side of the airfoil. On the pressure surface, the pressure is nearly equal to 60 percent of the length before the fluid is accelerated to the exit pressure condition. On suction surface, the fluid is accelerated in first 60 percent of the length to a low pressure and then decelerated to the exit pressure condition. The deceleration on the suction surface may lead to boundary layer separation. To avoid this, enough airfoils must be present.

12.3.2 Radial Variation

As the mass flow rate per unit area is higher in turbines, the blades are comparatively shorter and therefore little radial variation in aerodynamic properties takes place from hub to tip except in last few stages of the low-pressure turbine. Due to radial variation, the degree of reaction is almost zero at the hub and up to 40 percent at the tip.

12.4 TURBINE CHARACTERISTICS

Like compressor, for turbine also the following dimensionless parameters are involved:

$$\frac{p_{02}}{p_{01}}, \frac{T_{02}}{T_{01}}, \frac{m\sqrt{RT_{01}}}{D^2 p_{01}}, \frac{ND}{\sqrt{RT_{01}}}$$

For a particular machine the parameters R and D are fixed and, therefore, omitted and the dimensionless numbers which are not actually dimensionless, becomes

$$\frac{p_{02}}{p_{01}}, \frac{T_{02}}{T_{01}}, \frac{m\sqrt{T_{01}}}{p_{01}}, \frac{N}{\sqrt{T_{01}}}$$

If a turbine consists of more stages, the stage efficiency may be considered as equal to polytropic efficiency. Turbines have few stages, and it is preferable to work through turbine stage by stage. The overall efficiency, η_t is estimated from the ratio of actual to isentropic overall temperature drop.

Typical turbine characteristics plots are shown in figure 12.6. Suffix 3 and 4 are for inlet and outlet conditions of the turbine respectively. The plots are made between η_t and p_{03}/p_{04} for various values of $N/\sqrt{T_{03}}$ and also, between $m\sqrt{T_{03}}/p_{03}$ and p_{03}/p_{04} for various values of $N/\sqrt{T_{03}}$. The efficiency plot shows that η_t is reasonably constant over a wide range of rotational speed and pressure ratio. The maximum value of $m\sqrt{T_{03}}/p_{03}$ is reached at a pressure ratio which produces choking conditions at some point in the turbine. Normally, both the pressure ratio and the mass flow increase simultaneously as rotational speed is increased.

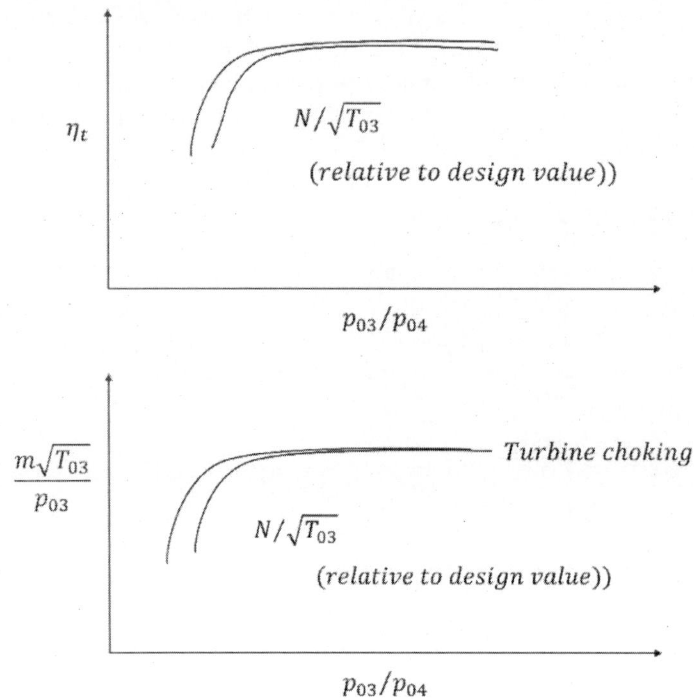

Figure 12.6: Typical Turbine Characteristics

12.5 BLADE MATERIALS AND COOLING

The fluid at high temperature and pressure enters a gas turbine through stationary airfoils, called stator blades or nozzles which accelerate the fluid, increasing its tangential momentum. Then it passes through rotating airfoils called rotor blades, that remove energy from the fluid by changing its tangential momentum. Successive pairs of nozzles and rotors remove additional energy from the fluid. The first stage nozzle blades are exposed to the highest turbine temperature. A blade can be defined as the medium of transfer of energy from the gases to the turbine. The first stage rotor blades are exposed to little lower temperature due to dilution of hot gases with first stage stator cooling air and relative velocity effects. The subsequent stages, both stator and rotor blades are exposed to lower temperatures.

Austenitic steel was used in early turbine blades during around 1940 which later turned to high nickel alloys. Subsequently, super alloys of nickel and cobalt were used. About 20% Cr (chromium) was used to protect against high temperature oxidation (corrosion). Later a part of chromium was replaced by aluminium and titanium. Titanium alloys are preferred over steel because of lower density (about 50%). Titanium also has a lower coefficient of thermal expansion which helps to reduce thermal stresses.

Blade cooling is the most effective way of maintaining high temperatures using the available material. In turbojet engines, the liquid systems have not proved to be practicable due to difficulties associated with channeling the liquid to and from the blades. The method used successfully in engines has been air cooling. With about 2% of the air mass flow used for cooling per blade row, the blade temperature can be reduced by 200–300 °C. This air is drawn from the main flow from the compressor before entering the turbine stage. The methods of external cooling are film cooling and transpiration or effusion cooling. Film cooling to local areas can be applied by drilling holes on precision cast blades by electric discharge machining, EDM. The cooling air flows out of these holes forms a thin film over the blade surfaces. Besides cooling the blade surface, it decreases the heat transfer from the hot gases to the blade material. This is shown in figure 12.7 below.

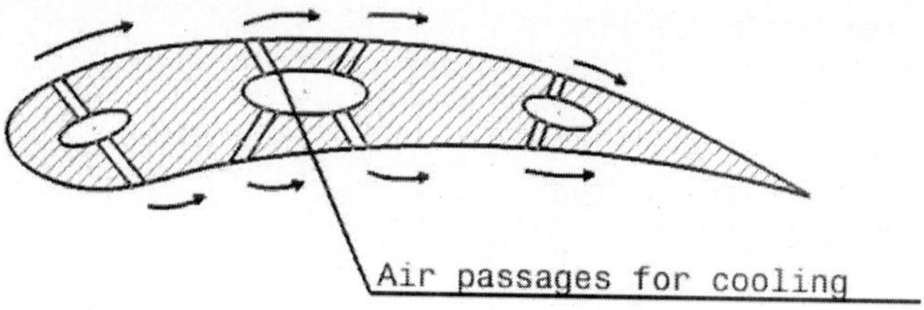

Figure 12.7: Film Cooling

In transpiration cooling, the air is allowed to effuse from the pores of the porous material as shown in Figure 12.8. This provides a blanket of cool air, insulating the metal of the turbine blade from the hot gas. Sintered wire composites are the most commonly used transpiring materials.

Internal cooling of blades is achieved by passing air through internal cooling passages from hub towards the blade tip. The cooling of the blade is by conduction and convection. Attempt has also been made to use ceramic materials in small turbine rotors and both silicon nitride and silicon carbide are developed. Ceramic blades are subject to brittle failure. Ceramic materials need extensive research and experimentation before it has an impact on gas turbine design.

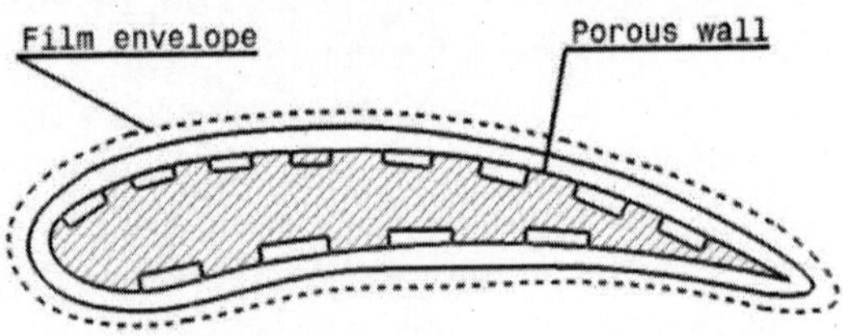

Figure 12.8: Transpiration Cooling

SUMMARY

The turbine is reverse of a compressor. The stress associated with turbine blades is higher because of higher work and temperature. The turbine handles products of combustion at very high temperature. In turbine, the stator comes first which is usually called a nozzle, followed by a rotor. In nozzle, the fluid is accelerated, and the static pressure is decreased. The net tangential component of velocity increases which produces torque on the rotor blades. The absolute velocity of fluid decreases in rotor. A set of a nozzle and rotor is called a stage.

The work output of a stage is given as,

$$\begin{aligned} W_S &= Uc_a(\tan\alpha_2 + \tan\alpha_3) \\ &= Uc_a(\tan\beta_2 + \tan\beta_3) \\ &= C_p.(\Delta T_{0S}) \\ &= C_p(T_{01} - T_{03}) \\ &= C_p(T_{02} - T_{03}) \end{aligned}$$

where, ΔT_{0S} is total temperature drop during stage.

Stage blade loading coefficient or temperature drop coefficient is defined as

$$\begin{aligned} \psi &= \frac{C_p(\Delta T_{0S})}{U^2} \\ &= \frac{C_a}{U}(\tan\beta_2 + \tan\beta_3) \end{aligned}$$

Degree of reaction, Λ

$$\Lambda = \frac{T_2 - T_3}{T_1 - T_3}$$

$$\Lambda = \frac{c_a}{2U}(\tan\beta_3 - \tan\beta_2)$$

Flow coefficient, ϕ

$$\phi = \frac{c_a}{U}$$

For 50% degree of reaction,

$$\beta_2 = \alpha_3 \text{ and } \beta_3 = \alpha_2$$

The efficiency plot shows that the turbine efficiency, η_t is reasonably constant over a wide range of rotational speed and pressure ratio. Normally, both the pressure ratio and mass flow increase simultaneously as the rotational speed is increased.

The titanium alloys are preferred as blade material than steel because of lower density. Titanium also has a lower coefficient of thermal expansion which helps to reduce the thermal stresses. Nickel alloys are also used as blade material. Blade cooling is the most effective way of maintaining high temperature with the available materials. About 2 percent of the air mass flow is used for cooling per blade row. The air is drawn from the main flow from the compressor before entering the turbine stage.

The design process is discussed in brief. The design is complex and needs background and knowledge of acceptable range of various parameters.

EXAMPLE

Example 12.1

In a single stage axial flow impulse turbine, the nozzle discharges the fluid onto the blades at an angle of 60° to the axial direction and the fluid leaves the blades with an absolute velocity of 350 m/s at an angle of 30° to the axial direction. The blades have equal inlet and outlet angles and there is no axial thrust, calculate the blade angle and the energy produced per unit mass flow of the fluid.

Solution

Refer Figure 12.9. Since the axial thrust is zero, c_a remains constant i.e.,

$$c_{a2} = c_{a3} = c_a$$

$$c_a = c_3 \cos \alpha_3$$
$$= 350 \times \cos 30$$
$$= 303.1\ m/s$$

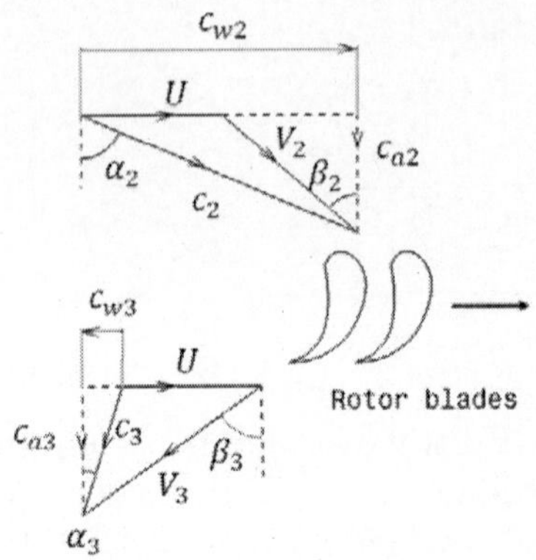

Figure 12.9: Velocity Triangles at Inlet and Outlet of Rotor

$$c_2 = \frac{c_a}{\cos \alpha_2}$$
$$= \frac{303.1}{\cos 60}$$
$$= 606.2\ m/s$$

Given, $\beta_2 = \beta_3 = \beta$

$$c_{w3} = c_a \tan \alpha_3$$
$$= 303.1 \times \tan 30$$
$$= 175\ m/s$$

$$c_{w2} = c_2 \sin \alpha_2$$
$$= 606.2 \times \sin 60$$
$$= 525\ m/s$$

from Figure 12.9,

$$c_{w2} = U + c_a \tan \beta$$
$$525 = U + 303.1 \times \tan \beta$$
$$U = 525 - (303.1\ \times \tan \beta)$$

also,

$$U + c_{w3} = c_a \tan \beta$$
$$U = 303.1\ \tan \beta - 175$$

Equate the two U values to get $\tan\beta$,

$$525 - (303.1 \times \tan\beta) = 303.1 \tan\beta - 175$$

$$525 + 175 = 606.2 \tan\beta$$

$$\tan\beta = \frac{700}{606.2} = 1.154$$

therefore,

$$\boldsymbol{\beta = 49.1^\circ}$$

now calculate U,

$$U = 303.1 \tan\beta - 175$$

$$= 303.1 \times 1.154 - 175$$

$$= 174.7\ m/s$$

calculate W_s,

$$W_s = (c_{w2} + c_{w3}) \times U$$

$$= (525 + 175) \times 174.7$$

$$= 122290\ J/kg$$

$$= \mathbf{122.29\ kJ/kg}$$

QUESTIONS

1. With the help of neat sketches, describe the functioning of a general axial flow turbine.
2. Draw the velocity triangles at nozzle inlet and rotor inlet and outlet for a single stage axial flow turbine and hence derive the expression for work output.
3. Define degree of reaction and discuss why 50% reaction is most suited. Also, draw the T-s diagram for the reaction stage of an axial flow turbine.
4. Discuss the various methods for cooling of blades in turbines.

Bibliography

[1] Anderson Jr, John D., Aircraft Performnace and Design, McGraw-Hill, 1999.

[2] Anderson Jr, John D., Introduction to Flight, Seventh Edition, McGraw-Hill, 2012.

[3] Bose, Tarit, Airbreathing propulsion—An Introduction, Springer, 2012.

[4] Bush, M.B., Normal Shock Tables, Retrieved January 7, 2023, fromhttps://web.iitd.ac.in/~pmvs/courses/mcl341/normal_shock.pdf.

[5] Farokhi Saeed, Aircraft Propulsion, Wiley, Second Edition, 2014.

[6] Ganesan V., Gas Turbines, 2nd Edition, Tata Mc Graw Hill Publishing company Limited, 2007.

[7] Ganesan V., Internal Combustion Engines, 4th Edition, Mc Graw Hill Education India Private Limited, New Delhi, 2013.

[8] Glassman I., Yetter Richard A., Combustion, Academic Press, Fourth Edition, 2008.

[9] Heywood, John B. Internal Combustion Engine Fundamentals, Mc Graw Hill Education (India) Private Limited, Chennai, 2017.

[10] Hill P.G. and Peterson C.R., Mechanics and Thermodynamics of Propulsion, Addison-Wesley Publishing Company, 1992.

[11] Holman J.P., Bhattacharyya S, Heat Transfer, Tenth Edition, Mc Graw Hill Education (India) Private Limited, New Delhi, 2018.

[12] Jouhari, A.K, Damodaran KA and Gupta, M.C., Development of a Jet Engine Combustion Chamber 7th National Conference on I.C. Engines and Combustion, KREC, Surathkal, 1981, pp. DS 1.01-1.05.

[13] Krishnan, Sumit, Jouhari, A.K. and Saxena Vinod, Experimental Investigations on Performance of a Propeller, 2016 (Unpublished work).

[14] Lefebvre, Arthur H. and Ballal, Dilip R., Gas Turbine Combustion, 3rd Edition, CRC Press, 2010.

[15] Mattingly, Jack D., Elements of Gas Turbine Propulsion, McGraw Hill Education, 2017.

[16] NACA Wartime Report, L-286, Originally Issued September 1943 as Advance Restricted Report 3125.

[17] Rathakrishnan, E.., Gas Dynamics, PHI Learning Private Limited, New Delhi, Second Edition, 200932.

[18] Ravi, P.A, Verma Vivek and Jouhari, A.K., Hydrogen Vehicle-Opportunities and Challenges, Eds. M. Kumar, et. al., Advances in Interdisciplinary Engineering, Lecture Notes in Mechanical Engineering, Springer, 2019.

[19] Roy, Bhaskar and Pradeep A.M., Jet Aircraft Propulsion, Lecture 26, IIT Bombay, NPTEL 2012.

[20] Saravanamuttoo HIH, Rogers GFC and Cohen H., Gas Turbine Theory, Pearson Education, 2005.

[21] Shapiro. A.H., Dynamics and Thermodynamics of Compressible Fluid Flow, Vols. I & II, Ronald Press, New York, 1953.

[22] Shevell Richard S, Fundamentals of Flight, Second Edition, Pearson, 1988.

[23] Sonntag Richard E., Borgnakke, C. and Van Wylen, Gordon J., Fundamentals of Thermodynamics, John Wiley & Sons, Inc., Sixth Edition, 2004.

[24] Srivastava, M.C. and Jouhari A.K., Composite Materials for Automobiles, in Proceedings, IC-RAMSD 2018, February 1-3, The M.S. University of Baroda, Eds. A.K. Vaish, *et al.*, 2018.

[25] Streeter, Victor L., Fluid Mechanics, McGraw Hill Book Company, INC, New York, 1962.

[26] Treager, I.E., Aircraft Gas Turbine Engine Technology, McGraw Hill, New York, Second Edition, 1979.

[27] Turns, Stephen R., Introduction to Combustion- Concepts and Applications, 3rd Edition McGraw Hill Education (India) Private Limited, Chennai, 2018.

[28] White, Frank M., Fluid Mechanics, 7th Edition, Tata McGraw Hill Education Private Limited, New Delhi, 2011.

Appendix A

Propeller chart for a four-bladed tractor propeller (blade activity factor 135).

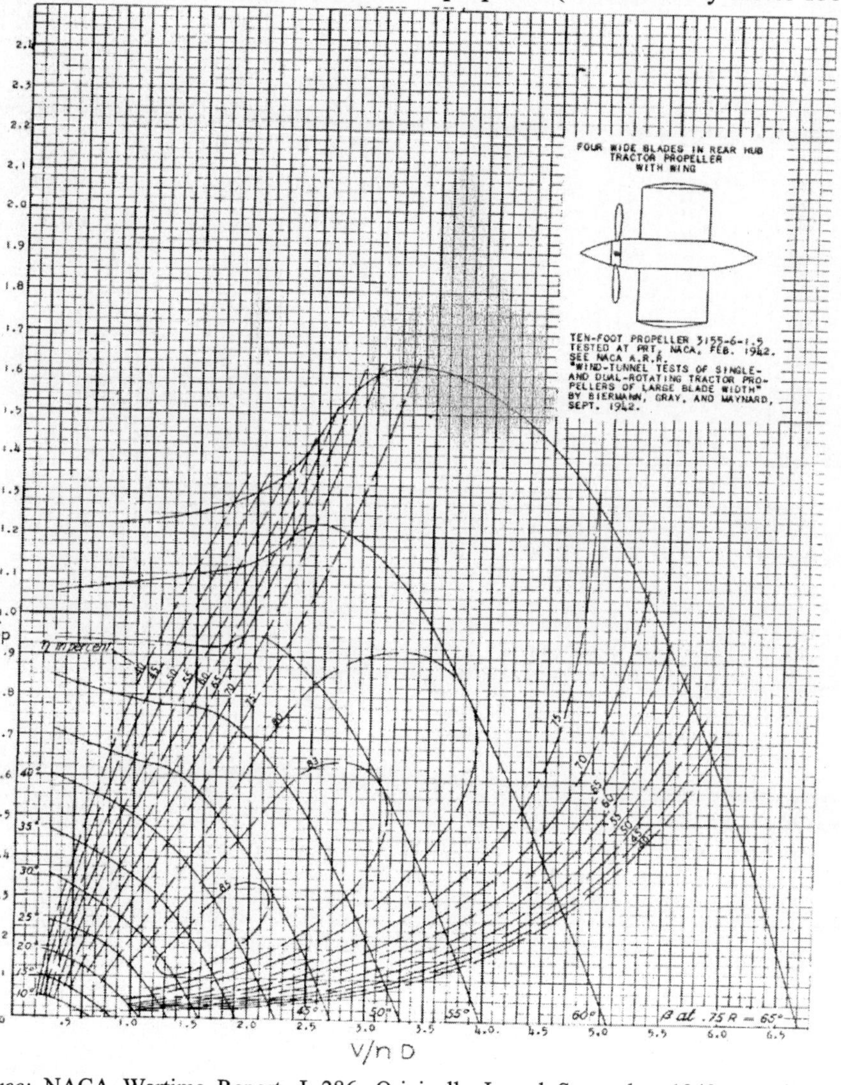

(*Source:* NACA Wartime Report, L-286, Originally Issued September 1943 as Advance Restricted Report 3125)

APPENDIX B

Normal Shock Tables ($\gamma = 1.4$)

M_1	M_2	P_2/P_1	ρ_2/ρ_1	T_2/T_1	P_{02}/P_{01}	P_1/P_{02}
1.00	1.0000	1.0000	1.0000	1.0000	1.0000	0.5283
1.01	0.9901	1.0235	1.0167	1.0066	1.0000	0.5221
1.02	0.9805	1.0471	1.0334	1.0132	1.0000	0.5160
1.03	0.9712	1.0711	1.0502	1.0198	1.0000	0.5100
1.04	0.9620	1.0952	1.0671	1.0263	0.9999	0.5039
1.05	0.9531	1.1196	1.0840	1.0328	0.9999	0.4979
1.06	0.9444	1.1442	1.1009	1.0393	0.9998	0.4920
1.07	0.9360	1.1691	1.1179	1.0458	0.9996	0.4861
1.08	0.9277	1.1941	1.1349	1.0522	0.9994	0.4803
1.09	0.9196	1.2195	1.1520	1.0586	0.9992	0.4746
1.10	0.9118	1.2450	1.1691	1.0649	0.9989	0.4689
1.11	0.9041	1.2708	1.1862	1.0713	0.9986	0.4632
1.12	0.8966	1.2968	1.2034	1.0776	0.9982	0.4576
1.13	0.8892	1.3231	1.2206	1.0840	0.9978	0.4521
1.14	0.8820	1.3495	1.2378	1.0903	0.9973	0.4467
1.15	0.8750	1.3763	1.2550	1.0966	0.9967	0.4413
1.16	0.8682	1.4032	1.2723	1.1029	0.9961	0.4360
1.17	0.8615	1.4304	1.2896	1.1092	0.9953	0.4307
1.18	0.8549	1.4578	1.3069	1.1154	0.9946	0.4255
1.19	0.8485	1.4855	1.3243	1.1217	0.9937	0.4204
1.20	0.8422	1.5133	1.3416	1.1280	0.9928	0.4154
1.21	0.8360	1.5415	1.3590	1.1343	0.9918	0.4104
1.22	0.8300	1.5698	1.3764	1.1405	0.9907	0.4055
1.23	0.8241	1.5984	1.3938	1.1468	0.9896	0.4006
1.24	0.8183	1.6272	1.4112	1.1531	0.9884	0.3958
1.25	0.8126	1.6563	1.4286	1.1594	0.9871	0.3911
1.26	0.8071	1.6855	1.4460	1.1657	0.9857	0.3865
1.27	0.8016	1.7151	1.4634	1.1720	0.9842	0.3819
1.28	0.7963	1.7448	1.4808	1.1783	0.9827	0.3774

M_1	M_2	P_2/P_1	ρ_2/ρ_1	T_2/T_1	P_{02}/P_{01}	P_1/P_{02}
1.29	0.7911	1.7748	1.4983	1.1846	0.9811	0.3729
1.30	0.7860	1.8050	1.5157	1.1909	0.9794	0.3685
1.31	0.7809	1.8355	1.5331	1.1972	0.9776	0.3642
1.32	0.7760	1.8661	1.5505	1.2035	0.9758	0.3599
1.33	0.7712	1.8971	1.5680	1.2099	0.9738	0.3557
1.34	0.7664	1.9282	1.5854	1.2162	0.9718	0.3516
1.35	0.7618	1.9596	1.6028	1.2226	0.9697	0.3475
1.36	0.7572	1.9912	1.6202	1.2290	0.9676	0.3435
1.37	0.7527	2.0231	1.6376	1.2354	0.9653	0.3395
1.38	0.7483	2.0551	1.6549	1.2418	0.9630	0.3356
1.39	0.7440	2.0875	1.6723	1.2482	0.9607	0.3317
1.40	0.7397	2.1200	1.6897	1.2547	0.9582	0.3280
1.41	0.7355	2.1528	1.7070	1.2612	0.9557	0.3242
1.42	0.7314	2.1858	1.7243	1.2676	0.9531	0.3205
1.43	0.7274	2.2191	1.7416	1.2741	0.9504	0.3169
1.44	0.7235	2.2525	1.7589	1.2807	0.9476	0.3133
1.45	0.7196	2.2863	1.7761	1.2872	0.9448	0.3098
1.46	0.7157	2.3202	1.7934	1.2938	0.9420	0.3063
1.47	0.7120	2.3544	1.8106	1.3003	0.9390	0.3029
1.48	0.7083	2.3888	1.8278	1.3069	0.9360	0.2996
1.49	0.7047	2.4235	1.8449	1.3136	0.9329	0.2962
1.50	0.7011	2.4583	1.8621	1.3202	0.9298	0.2930
1.51	0.6976	2.4935	1.8792	1.3269	0.9266	0.2898
1.52	0.6941	2.5288	1.8963	1.3336	0.9233	0.2866
1.53	0.6907	2.5644	1.9133	1.3403	0.9200	0.2835
1.54	0.6874	2.6002	1.9303	1.3470	0.9166	0.2804
1.55	0.6841	2.6363	1.9473	1.3538	0.9132	0.2773
1.56	0.6809	2.6725	1.9643	1.3606	0.9097	0.2743
1.57	0.6777	2.7091	1.9812	1.3674	0.9062	0.2714
1.58	0.6746	2.7458	1.9981	1.3742	0.9026	0.2685
1.59	0.6715	2.7828	2.0149	1.3811	0.8989	0.2656
1.60	0.6684	2.8200	2.0317	1.3880	0.8952	0.2628

M_1	M_2	P_2/P_1	ρ_2/ρ_1	T_2/T_1	P_{02}/P_{01}	P_1/P_{02}
1.61	0.6655	2.8575	2.0485	1.3949	0.8915	0.2600
1.62	0.6625	2.8951	2.0653	1.4018	0.8877	0.2573
1.63	0.6596	2.9331	2.0820	1.4088	0.8838	0.2546
1.64	0.6568	2.9712	2.0986	1.4158	0.8799	0.2519
1.65	0.6540	3.0096	2.1152	1.4228	0.8760	0.2493
1.66	0.6512	3.0482	2.1318	1.4299	0.8720	0.2467
1.67	0.6485	3.0871	2.1484	1.4369	0.8680	0.2442
1.68	0.6458	3.1261	2.1649	1.4440	0.8639	0.2417
1.69	0.6431	3.1655	2.1813	1.4512	0.8599	0.2392
1.70	0.6405	3.2050	2.1977	1.4583	0.8557	0.2368
1.71	0.6380	3.2448	2.2141	1.4655	0.8516	0.2343
1.72	0.6355	3.2848	2.2304	1.4727	0.8474	0.2320
1.73	0.6330	3.3251	2.2467	1.4800	0.8431	0.2296
1.74	0.6305	3.3655	2.2629	1.4873	0.8389	0.2273
1.75	0.6281	3.4063	2.2791	1.4946	0.8346	0.2251
1.76	0.6257	3.4472	2.2952	1.5019	0.8302	0.2228
1.77	0.6234	3.4884	2.3113	1.5093	0.8259	0.2206
1.78	0.6210	3.5298	2.3273	1.5167	0.8215	0.2184
1.79	0.6188	3.5715	2.3433	1.5241	0.8171	0.2163
1.80	0.6165	3.6133	2.3592	1.5316	0.8127	0.2142
1.81	0.6143	3.6555	2.3751	1.5391	0.8082	0.2121
1.82	0.6121	3.6978	2.3909	1.5466	0.8038	0.2100
1.83	0.6099	3.7404	2.4067	1.5541	0.7993	0.2080
1.84	0.6078	3.7832	2.4224	1.5617	0.7948	0.2060
1.85	0.6057	3.8263	2.4381	1.5693	0.7902	0.2040
1.86	0.6036	3.8695	2.4537	1.5770	0.7857	0.2020
1.87	0.6016	3.9131	2.4693	1.5847	0.7811	0.2001
1.88	0.5996	3.9568	2.4848	1.5924	0.7765	0.1982
1.89	0.5976	4.0008	2.5003	1.6001	0.7720	0.1963
1.90	0.5956	4.0450	2.5157	1.6079	0.7674	0.1945
1.91	0.5937	4.0895	2.5310	1.6157	0.7627	0.1927
1.92	0.5918	4.1341	2.5463	1.6236	0.7581	0.1909

M_1	M_2	P_2/P_1	ρ_2/ρ_1	T_2/T_1	P_{02}/P_{01}	P_1/P_{02}
1.93	0.5899	4.1791	2.5616	1.6314	0.7535	0.1891
1.94	0.5880	4.2242	2.5767	1.6394	0.7488	0.1873
1.95	0.5862	4.2696	2.5919	1.6473	0.7442	0.1856
1.96	0.5844	4.3152	2.6069	1.6553	0.7395	0.1839
1.97	0.5826	4.3611	2.6220	1.6633	0.7349	0.1822
1.98	0.5808	4.4071	2.6369	1.6713	0.7302	0.1806
1.99	0.5791	4.4535	2.6518	1.6794	0.7255	0.1789
2.00	0.5774	4.5000	2.6667	1.6875	0.7209	0.1773
2.01	0.5757	4.5468	2.6815	1.6956	0.7162	0.1757
2.02	0.5740	4.5938	2.6962	1.7038	0.7115	0.1741
2.03	0.5723	4.6411	2.7109	1.7120	0.7069	0.1726
2.04	0.5707	4.6885	2.7255	1.7203	0.7022	0.1710
2.05	0.5691	4.7363	2.7400	1.7285	0.6975	0.1695
2.06	0.5675	4.7842	2.7545	1.7369	0.6928	0.1680
2.07	0.5659	4.8324	2.7689	1.7452	0.6882	0.1665
2.08	0.5643	4.8808	2.7833	1.7536	0.6835	0.1651
2.09	0.5628	4.9294	2.7976	1.7620	0.6789	0.1636
2.10	0.5613	4.9783	2.8119	1.7705	0.6742	0.1622
2.11	0.5598	5.0274	2.8261	1.7789	0.6696	0.1608
2.12	0.5583	5.0768	2.8402	1.7875	0.6649	0.1594
2.13	0.5568	5.1264	2.8543	1.7960	0.6603	0.1580
2.14	0.5554	5.1762	2.8683	1.8046	0.6557	0.1567
2.15	0.5540	5.2262	2.8823	1.8132	0.6511	0.1553
2.16	0.5525	5.2765	2.8962	1.8219	0.6464	0.1540
2.17	0.5511	5.3270	2.9101	1.8306	0.6419	0.1527
2.18	0.5498	5.3778	2.9238	1.8393	0.6373	0.1514
2.19	0.5484	5.4288	2.9376	1.8481	0.6327	0.1501
2.20	0.5471	5.4800	2.9512	1.8569	0.6281	0.1489
2.21	0.5457	5.5314	2.9648	1.8657	0.6236	0.1476
2.22	0.5444	5.5831	2.9784	1.8746	0.6191	0.1464
2.23	0.5431	5.6350	2.9918	1.8835	0.6145	0.1452
2.24	0.5418	5.6872	3.0053	1.8924	0.6100	0.1440

M_1	M_2	P_2/P_1	ρ_2/ρ_1	T_2/T_1	P_{02}/P_{01}	P_1/P_{02}
2.25	0.5406	5.7396	3.0186	1.9014	0.6055	0.1428
2.26	0.5393	5.7922	3.0319	1.9104	0.6011	0.1416
2.27	0.5381	5.8450	3.0452	1.9194	0.5966	0.1405
2.28	0.5368	5.8981	3.0584	1.9285	0.5921	0.1394
2.29	0.5356	5.9514	3.0715	1.9376	0.5877	0.1382
2.30	0.5344	6.0050	3.0845	1.9468	0.5833	0.1371
2.31	0.5332	6.0588	3.0976	1.9560	0.5789	0.1360
2.32	0.5321	6.1128	3.1105	1.9652	0.5745	0.1349
2.33	0.5309	6.1670	3.1234	1.9745	0.5702	0.1338
2.34	0.5297	6.2215	3.1362	1.9838	0.5658	0.1328
2.35	0.5286	6.2762	3.1490	1.9931	0.5615	0.1317
2.36	0.5275	6.3312	3.1617	2.0025	0.5572	0.1307
2.37	0.5264	6.3864	3.1743	2.0119	0.5529	0.1296
2.38	0.5253	6.4418	3.1869	2.0213	0.5486	0.1286
2.39	0.5242	6.4974	3.1994	2.0308	0.5444	0.1276
2.40	0.5231	6.5533	3.2119	2.0403	0.5401	0.1266
2.41	0.5221	6.6095	3.2243	2.0499	0.5359	0.1256
2.42	0.5210	6.6658	3.2367	2.0595	0.5317	0.1247
2.43	0.5200	6.7224	3.2489	2.0691	0.5276	0.1237
2.44	0.5189	6.7792	3.2612	2.0788	0.5234	0.1228
2.45	0.5179	6.8362	3.2733	2.0885	0.5193	0.1218
2.46	0.5169	6.8935	3.2855	2.0982	0.5152	0.1209
2.47	0.5159	6.9510	3.2975	2.1080	0.5111	0.1200
2.48	0.5149	7.0088	3.3095	2.1178	0.5071	0.1191
2.49	0.5140	7.0668	3.3215	2.1276	0.5030	0.1182
2.50	0.5130	7.1250	3.3333	2.1375	0.4990	0.1173
2.51	0.5120	7.1834	3.3452	2.1474	0.4950	0.1164
2.52	0.5111	7.2421	3.3569	2.1574	0.4911	0.1155
2.53	0.5102	7.3010	3.3686	2.1674	0.4871	0.1147
2.54	0.5092	7.3602	3.3803	2.1774	0.4832	0.1138
2.55	0.5083	7.4196	3.3919	2.1875	0.4793	0.1130
2.56	0.5074	7.4792	3.4034	2.1976	0.4754	0.1122

M_1	M_2	P_2/P_1	ρ_2/ρ_1	T_2/T_1	P_{02}/P_{01}	P_1/P_{02}
2.57	0.5065	7.5390	3.4149	2.2077	0.4715	0.1113
2.58	0.5056	7.5991	3.4263	2.2179	0.4677	0.1105
2.59	0.5047	7.6594	3.4377	2.2281	0.4639	0.1097
2.60	0.5039	7.7200	3.4490	2.2383	0.4601	0.1089
2.61	0.5030	7.7808	3.4602	2.2486	0.4564	0.1081
2.62	0.5022	7.8418	3.4714	2.2590	0.4526	0.1073
2.63	0.5013	7.9030	3.4826	2.2693	0.4489	0.1066
2.64	0.5005	7.9645	3.4937	2.2797	0.4452	0.1058
2.65	0.4996	8.0262	3.5047	2.2902	0.4416	0.1051
2.66	0.4988	8.0882	3.5157	2.3006	0.4379	0.1043
2.67	0.4980	8.1504	3.5266	2.3111	0.4343	0.1036
2.68	0.4972	8.2128	3.5374	2.3217	0.4307	0.1028
2.69	0.4964	8.2754	3.5482	2.3323	0.4271	0.1021
2.70	0.4956	8.3383	3.5590	2.3429	0.4236	0.1014
2.71	0.4949	8.4014	3.5697	2.3536	0.4201	0.1007
2.72	0.4941	8.4648	3.5803	2.3642	0.4166	0.1000
2.73	0.4933	8.5284	3.5909	2.3750	0.4131	.9929 E-1
2.74	0.4926	8.5922	3.6015	2.3858	0.4097	.9860 E-1
2.75	0.4918	8.6563	3.6119	2.3966	0.4062	9792 E-1
2.76	0.4911	8.7205	3.6224	2.4074	0.4028	.9724 E-1
2.77	0.4903	8.7851	3.6327	2.4183	0.3994	.9658 E-1
2.78	0.4896	8.8498	3.6431	2.4292	0.3961	.9591 E-1
2.79	0.4889	8.9148	3.6533	2.4402	0.3928	.9526 E-1
2.80	0.4882	8.9800	3.6636	2.4512	0.3895	.9461 E-1
2.81	0.4875	9.0454	3.6737	2.4622	0.3862	.9397 E-1
2.82	0.4868	9.1111	3.6838	2.4733	0.3829	.9334 E-1
2.83	0.4861	9.1770	3.6939	2.4844	0.3797	.9271 E-1
2.84	0.4854	9.2432	3.7039	2.4955	0.3765	.9209 E-1
2.85	0.4847	9.3096	3.7139	2.5067	0.3733	.9147 E-1
2.86	0.4840	9.3762	3.7238	2.5179	0.3701	.9086 E-1
2.87	0.4833	9.4430	3.7336	2.5292	0.3670	.9026 E-1
2.88	0.4827	9.5101	3.7434	2.5405	0.3639	.8966 E-1

M_1	M_2	P_2/P_1	ρ_2/ρ_1	T_2/T_1	P_{02}/P_{01}	P_1/P_{02}
2.89	0.4820	9.5774	3.7532	2.5518	0.3608	.8906 E-1
2.90	0.4814	9.6450	3.7629	2.5632	0.3577	.8848 E-1
2.91	0.4807	9.7128	3.7725	2.5746	0.3547	.8790 E-1
2.92	0.4801	9.7808	3.7821	2.5861	0.3517	.8732 E-1
2.93	0.4795	9.8490	3.7917	2.5976	0.3487	.8675 E-1
2.94	0.4788	9.9175	3.8012	2.6091	0.3457	.8619 E-1
2.95	0.4782	9.9862	3.8106	2.6206	0.3428	.8563 E-1
2.96	0.4776	10.0552	3.8200	2.6322	0.3398	.8507 E-1
2.97	0.4770	10.1244	3.8294	2.6439	0.3369	.8453 E-1
2.98	0.4764	10.1938	3.8387	2.6555	0.3340	.8398 E-1
2.99	0.4758	10.2635	3.8479	2.6673	0.3312	.8345 E-1
3.00	0.4752	10.3333	3.8571	2.6790	0.3283	.8291 E-1
3.01	0.4746	10.4035	3.8663	2.6908	0.3255	.8238 E-1
3.02	0.4740	10.4738	3.8754	2.7026	0.3227	.8186 E-1
3.03	0.4734	10.5444	3.8845	2.7145	0.3200	.8134 E-1
3.04	0.4729	10.6152	3.8935	2.7264	0.3172	.8083 E-1
3.05	0.4723	10.6863	3.9025	2.7383	0.3145	.8032 E-1
3.06	0.4717	10.7575	3.9114	2.7503	0.3118	.7982 E-1
3.07	0.4712	10.8291	3.9203	2.7623	0.3091	.7932 E-1
3.08	0.4706	10.9008	3.9291	2.7744	0.3065	.7882 E-1
3.09	0.4701	10.9728	3.9379	2.7865	0.3038	.7833 E-1
3.10	0.4695	11.0450	3.9466	2.7986	0.3012	.7785 E-1
3.11	0.4690	11.1175	3.9553	2.8108	0.2986	.7737 E-1
3.12	0.4685	11.1901	3.9639	2.8230	0.2960	.7689 E-1
3.13	0.4679	11.2631	3.9725	2.8352	0.2935	.7642 E-1
3.14	0.4674	11.3362	3.9811	2.8475	0.2910	.7595 E-1
3.15	0.4669	11.4096	3.9896	2.8598	0.2885	.7549 E-1
3.16	0.4664	11.4832	3.9981	2.8722	0.2860	.7503 E-1
3.17	0.4659	11.5571	4.0065	2.8846	0.2835	.7457 E-1
3.18	0.4654	11.6311	4.0149	2.8970	0.2811	.7412 E-1
3.19	0.4648	11.7055	4.0232	2.9095	0.2786	.7367 E-1
3.20	0.4643	11.7800	4.0315	2.9220	0.2762	.7323 E-1

M_1	M_2	P_2/P_1	ρ_2/ρ_1	T_2/T_1	P_{02}/P_{01}	P_1/P_{02}
3.21	0.4639	11.8548	4.0397	2.9345	0.2738	.7279 E-1
3.22	0.4634	11.9298	4.0479	2.9471	0.2715	.7235 E-1
3.23	0.4629	12.0051	4.0561	2.9598	0.2691	.7192 E-1
3.24	0.4624	12.0805	4.0642	2.9724	0.2668	.7149 E-1
3.25	0.4619	12.1563	4.0723	2.9851	0.2645	.7107 E-1
3.26	0.4614	12.2322	4.0803	2.9979	0.2622	.7065 E-1
3.27	0.4610	12.3084	4.0883	3.0106	0.2600	.7023 E-1
3.28	0.4605	12.3848	4.0963	3.0234	0.2577	.6982 E-1
3.29	0.4600	12.4615	4.1042	3.0363	0.2555	.6941 E-1
3.30	0.4596	12.5383	4.1120	3.0492	0.2533	.6900 E-1
3.31	0.4591	12.6155	4.1198	3.0621	0.2511	.6860 E-1
3.32	0.4587	12.6928	4.1276	3.0751	0.2489	.6820 E-1
3.33	0.4582	12.7704	4.1354	3.0881	0.2468	.6781 E-1
3.34	0.4578	12.8482	4.1431	3.1011	0.2446	.6741 E-1
3.35	0.4573	12.9263	4.1507	3.1142	0.2425	.6702 E-1
3.36	0.4569	13.0045	4.1583	3.1273	0.2404	.6664 E-1
3.37	0.4565	13.0831	4.1659	3.1405	0.2383	.6626 E-1
3.38	0.4560	13.1618	4.1734	3.1537	0.2363	.6588 E-1
3.39	0.4556	13.2408	4.1809	3.1669	0.2342	.6550 E-1
3.40	0.4552	13.3200	4.1884	3.1802	0.2322	.6513 E-1
3.41	0.4548	13.3994	4.1958	3.1935	0.2302	.6476 E-1
3.42	0.4544	13.4791	4.2032	3.2069	0.2282	.6439 E-1
3.43	0.4540	13.5590	4.2105	3.2203	0.2263	.6403 E-1
3.44	0.4535	13.6392	4.2179	3.2337	0.2243	.6367 E-1
3.45	0.4531	13.7196	4.2251	3.2472	0.2224	.6331 E-1
3.46	0.4527	13.8002	4.2323	3.2607	0.2205	.6296 E-1
3.47	0.4523	13.8811	4.2395	3.2742	0.2186	.6261 E-1
3.48	0.4519	13.9621	4.2467	3.2878	0.2167	.6226 E-1
3.49	0.4515	14.0435	4.2538	3.3014	0.2148	.6191 E-1
3.50	0.4512	14.1250	4.2609	3.3151	0.2129	.6157 E-1
3.51	0.4508	14.2068	4.2679	3.3287	0.2111	.6123 E-1
3.52	0.4504	14.2888	4.2749	3.3425	0.2093	.6089 E-1

M_1	M_2	P_2/P_1	ρ_2/ρ_1	T_2/T_1	P_{02}/P_{01}	P_1/P_{02}
3.53	0.4500	14.3711	4.2819	3.3563	0.2075	.6056 E-1
3.54	0.4496	14.4535	4.2888	3.3701	0.2057	.6023 E-1
3.55	0.4492	14.5363	4.2957	3.3839	0.2039	.5990 E-1
3.56	0.4489	14.6192	4.3026	3.3978	0.2022	.5957 E-1
3.57	0.4485	14.7024	4.3094	3.4117	0.2004	.5925 E-1
3.58	0.4481	14.7858	4.3162	3.4257	0.1987	.5892 E-1
3.59	0.4478	14.8695	4.3229	3.4397	0.1970	.5861 E-1
3.60	0.4474	14.9533	4.3296	3.4537	0.1953	.5829 E-1
3.61	0.4471	15.0375	4.3363	3.4678	0.1936	.5798 E-1
3.62	0.4467	15.1218	4.3429	3.4819	0.1920	.5767 E-1
3.63	0.4463	15.2064	4.3496	3.4961	0.1903	.5736 E-1
3.64	0.4460	15.2912	4.3561	3.5103	0.1887	.5705 E-1
3.65	0.4456	15.3763	4.3627	3.5245	0.1871	.5675 E-1
3.66	0.4453	15.4615	4.3692	3.5388	0.1855	.5645 E-1
3.67	0.4450	15.5471	4.3756	3.5531	0.1839	.5615 E-1
3.68	0.4446	15.6328	4.3821	3.5674	0.1823	.5585 E-1
3.69	0.4443	15.7188	4.3885	3.5818	0.1807	.5556 E-1
3.70	0.4439	15.8050	4.3949	3.5962	0.1792	.5526 E-1
3.71	0.4436	15.8915	4.4012	3.6107	0.1777	.5497 E-1
3.72	0.4433	15.9781	4.4075	3.6252	0.1761	.5469 E-1
3.73	0.4430	16.0650	4.4138	3.6397	0.1746	.5440 E-1
3.74	0.4426	16.1522	4.4200	3.6543	0.1731	.5412 E-1
3.75	0.4423	16.2396	4.4262	3.6689	0.1717	.5384 E-1
3.76	0.4420	16.3272	4.4324	3.6836	0.1702	.5356 E-1
3.77	0.4417	16.4150	4.4385	3.6983	0.1687	.5328 E-1
3.78	0.4414	16.5031	4.4447	3.7130	0.1673	.5301 E-1
3.79	0.4410	16.5914	4.4507	3.7278	0.1659	.5274 E-1
3.80	0.4407	16.6800	4.4568	3.7426	0.1645	.5247 E-1
3.81	0.4404	16.7688	4.4628	3.7575	0.1631	.5220 E-1
3.82	0.4401	16.8578	4.4688	3.7723	0.1617	.5193 E-1
3.83	0.4398	16.9471	4.4747	3.7873	0.1603	.5167 E-1
3.84	0.4395	17.0365	4.4807	3.8022	0.1589	.5140 E-1

M_1	M_2	P_2/P_1	ρ_2/ρ_1	T_2/T_1	P_{02}/P_{01}	P_1/P_{02}
3.85	0.4392	17.1263	4.4866	3.8172	0.1576	.5114 E-1
3.86	0.4389	17.2162	4.4924	3.8323	0.1563	.5089 E-1
3.87	0.4386	17.3064	4.4983	3.8473	0.1549	.5063 E-1
3.88	0.4383	17.3968	4.5041	3.8625	0.1536	.5038 E-1
3.89	0.4380	17.4875	4.5098	3.8776	0.1523	.5012 E-1
3.90	0.4377	17.5783	4.5156	3.8928	0.1510	.4987 E-1
3.91	0.4375	17.6695	4.5213	3.9080	0.1497	.4962 E-1
3.92	0.4372	17.7608	4.5270	3.9233	0.1485	.4938 E-1
3.93	0.4369	17.8524	4.5326	3.9386	0.1472	.4913 E-1
3.94	0.4366	17.9442	4.5383	3.9540	0.1460	.4889 E-1
3.95	0.4363	18.0363	4.5439	3.9694	0.1448	.4865 E-1
3.96	0.4360	18.1285	4.5494	3.9848	0.1435	.4841 E-1
3.97	0.4358	18.2211	4.5550	4.0003	0.1423	.4817 E-1
3.98	0.4355	18.3138	4.5605	4.0158	0.1411	.4793 E-1
3.99	0.4352	18.4068	4.5660	4.0313	0.1399	.4770 E-1
4.00	0.4350	18.5000	4.5714	4.0469	0.1388	.4747 E-1
4.01	0.4347	18.5935	4.5769	4.0625	0.1376	.4723 E-1
4.02	0.4344	18.6871	4.5823	4.0782	0.1364	.4700 E-1
4.03	0.4342	18.7811	4.5876	4.0938	0.1353	.4678 E-1
4.04	0.4339	18.8752	4.5930	4.1096	0.1342	.4655 E-1
4.05	0.4336	18.9696	4.5983	4.1254	0.1330	.4633 E-1
4.06	0.4334	19.0642	4.6036	4.1412	0.1319	.4610 E-1
4.07	0.4331	19.1590	4.6089	4.1570	0.1308	.4588 E-1
4.08	0.4329	19.2541	4.6141	4.1729	0.1297	.4566 E-1
4.09	0.4326	19.3494	4.6193	4.1888	0.1286	.4544 E-1
4.10	0.4324	19.4450	4.6245	4.2048	0.1276	.4523 E-1
4.11	0.4321	19.5408	4.6296	4.2208	0.1265	.4501 E-1
4.12	0.4319	19.6368	4.6348	4.2368	0.1254	.4480 E-1
4.13	0.4316	19.7330	4.6399	4.2529	0.1244	.4459 E-1
4.14	0.4314	19.8295	4.6450	4.2690	0.1234	.4438 E-1
4.15	0.4311	19.9263	4.6500	4.2852	0.1223	.4417 E-1
4.16	0.4309	20.0232	4.6550	4.3014	0.1213	.4396 E-1

M_1	M_2	P_2/P_1	ρ_2/ρ_1	T_2/T_1	P_{02}/P_{01}	P_1/P_{02}
4.17	0.4306	20.1204	4.6601	4.3176	0.1203	.4375 E-1
4.18	0.4304	20.2178	4.6650	4.3339	0.1193	.4355 E-1
4.19	0.4302	20.3155	4.6700	4.3502	0.1183	.4334 E-1
4.20	0.4299	20.4133	4.6749	4.3666	0.1173	.4314 E-1
4.21	0.4297	20.5115	4.6798	4.3830	0.1164	.4294 E-1
4.22	0.4295	20.6098	4.6847	4.3994	0.1154	.4274 E-1
4.23	0.4292	20.7084	4.6896	4.4159	0.1144	.4255 E-1
4.24	0.4290	20.8072	4.6944	4.4324	0.1135	.4235 E-1
4.25	0.4288	20.9063	4.6992	4.4489	0.1126	.4215 E-1
4.26	0.4286	21.0055	4.7040	4.4655	0.1116	.4196 E-1
4.27	0.4283	21.1051	4.7087	4.4821	0.1107	.4177 E-1
4.28	0.4281	21.2048	4.7135	4.4988	0.1098	.4158 E-1
4.29	0.4279	21.3048	4.7182	4.5155	0.1089	.4139 E-1
4.30	0.4277	21.4050	4.7229	4.5322	0.1080	.4120 E-1
4.31	0.4275	21.5055	4.7275	4.5490	0.1071	.4101 E-1
4.32	0.4272	21.6061	4.7322	4.5658	0.1062	.4082 E-1
4.33	0.4270	21.7071	4.7368	4.5827	0.1054	.4064 E-1
4.34	0.4268	21.8082	4.7414	4.5995	0.1045	.4046 E-1
4.35	0.4266	21.9096	4.7460	4.6165	0.1036	.4027 E-1
4.36	0.4264	22.0112	4.7505	4.6334	0.1028	.4009 E-1
4.37	0.4262	22.1131	4.7550	4.6505	0.1020	.3991 E-1
4.38	0.4260	22.2151	4.7595	4.6675	0.1011	.3973 E-1
4.39	0.4258	22.3174	4.7640	4.6846	0.1003	.3956 E-1
4.40	0.4255	22.4200	4.7685	4.7017	.9948 E-1	.3938 E-1
4.41	0.4253	22.5228	4.7729	4.7189	.9867 E-1	.3921 E-1
4.42	0.4251	22.6258	4.7773	4.7361	.9787 E-1	.3903 E-1
4.43	0.4249	22.7290	4.7817	4.7533	.9707 E-1	.3886 E-1
4.44	0.4247	22.8325	4.7861	4.7706	.9628 E-1	.3869 E-1
4.45	0.4245	22.9362	4.7904	4.7879	.9550 E-1	.3852 E-1
4.46	0.4243	23.0402	4.7948	4.8053	.9473 E-1	.3835 E-1
4.47	0.4241	23.1444	4.7991	4.8227	.9396 E-1	.3818 E-1
4.48	0.4239	23.2488	4.8034	4.8401	.9320 E-1	.3801 E-1

M_1	M_2	P_2/P_1	ρ_2/ρ_1	T_2/T_1	P_{02}/P_{01}	P_1/P_{02}
4.49	0.4237	23.3534	4.8076	4.8576	.9244 E-1	.3785 E-1
4.50	0.4236	23.4583	4.8119	4.8751	.9170 E-1	.3768 E-1
4.51	0.4234	23.5635	4.8161	4.8926	.9096 E-1	.3752 E-1
4.52	0.4232	23.6688	4.8203	4.9102	.9022 E-1	.3735 E-1
4.53	0.4230	23.7744	4.8245	4.9279	.8950 E-1	.3719 E-1
4.54	0.4228	23.8802	4.8287	4.9455	.8878 E-1	.3703 E-1
4.55	0.4226	23.9863	4.8328	4.9632	.8806 E-1	.3687 E-1
4.56	0.4224	24.0925	4.8369	4.9810	.8735 E-1	.3671 E-1
4.57	0.4222	24.1991	4.8410	4.9987	.8665 E-1	.3656 E-1
4.58	0.4220	24.3058	4.8451	5.0166	.8596 E-1	.3640 E-1
4.59	0.4219	24.4128	4.8492	5.0344	.8527 E-1	.3624 E-1
4.60	0.4217	24.5200	4.8532	5.0523	.8459 E-1	.3609 E-1
4.61	0.4215	24.6275	4.8572	5.0703	.8391 E-1	.3593 E-1
4.62	0.4213	24.7351	4.8612	5.0882	.8324 E-1	.3578 E-1
4.63	0.4211	24.8431	4.8652	5.1063	.8257 E-1	.3563 E-1
4.64	0.4210	24.9512	4.8692	5.1243	.8192 E-1	.3548 E-1
4.65	0.4208	25.0596	4.8731	5.1424	.8126 E-1	.3533 E-1
4.66	0.4206	25.1682	4.8771	5.1605	.8062 E-1	.3518 E-1
4.67	0.4204	25.2771	4.8810	5.1787	.7998 E-1	.3503 E-1
4.68	0.4203	25.3861	4.8849	5.1969	.7934 E-1	.3488 E-1
4.69	0.4201	25.4955	4.8887	5.2151	.7871 E-1	.3474 E-1
4.70	0.4199	25.6050	4.8926	5.2334	.7809 E-1	.3459 E-1
4.71	0.4197	25.7148	4.8964	5.2518	.7747 E-1	.3445 E-1
4.72	0.4196	25.8248	4.9002	5.2701	.7685 E-1	.3431 E-1
4.73	0.4194	25.9350	4.9040	5.2885	.7625 E-1	.3416 E-1
4.74	0.4192	26.0455	4.9078	5.3070	.7564 E-1	.3402 E-1
4.75	0.4191	26.1562	4.9116	5.3254	.7505 E-1	.3388 E-1
4.76	0.4189	26.2672	4.9153	5.3440	.7445 E-1	.3374 E-1
4.77	0.4187	26.3784	4.9190	5.3625	.7387 E-1	.3360 E-1
4.78	0.4186	26.4898	4.9227	5.3811	.7329 E-1	.3346 E-1
4.79	0.4184	26.6014	4.9264	5.3997	.7271 E-1	.3333 E-1
4.80	0.4183	26.7133	4.9301	5.4184	.7214 E-1	.3319 E-1

M_1	M_2	P_2/P_1	ρ_2/ρ_1	T_2/T_1	P_{02}/P_{01}	P_1/P_{02}
4.81	0.4181	26.8254	4.9338	5.4371	.7157 E-1	.3305 E-1
4.82	0.4179	26.9378	4.9374	5.4559	.7101 E-1	.3292 E-1
4.83	0.4178	27.0504	4.9410	5.4747	.7046 E-1	.3278 E-1
4.84	0.4176	27.1632	4.9446	5.4935	.6991 E-1	.3265 E-1
4.85	0.4175	27.2763	4.9482	5.5124	.6936 E-1	.3252 E-1
4.86	0.4173	27.3895	4.9518	5.5313	.6882 E-1	.3239 E-1
4.87	0.4172	27.5031	4.9553	5.5502	.6828 E-1	.3226 E-1
4.88	0.4170	27.6168	4.9589	5.5692	.6775 E-1	.3213 E-1
4.89	0.4169	27.7308	4.9624	5.5882	.6722 E-1	.3200 E-1
4.90	0.4167	27.8450	4.9659	5.6073	.6670 E-1	.3187 E-1
4.91	0.4165	27.9595	4.9694	5.6264	.6618 E-1	.3174 E-1
4.92	0.4164	28.0741	4.9728	5.6455	.6567 E-1	.3161 E-1
4.93	0.4163	28.1891	4.9763	5.6647	.6516 E-1	.3149 E-1
4.94	0.4161	28.3042	4.9797	5.6839	.6465 E-1	.3136 E-1
4.95	0.4160	28.4196	4.9831	5.7032	.6415 E-1	.3124 E-1
4.96	0.4158	28.5352	4.9865	5.7224	.6366 E-1	.3111 E-1
4.97	0.4157	28.6511	4.9899	5.7418	.6317 E-1	.3099 E-1
4.98	0.4155	28.7671	4.9933	5.7611	.6268 E-1	.3087 E-1
4.99	0.4154	28.8835	4.9967	5.7806	.6220 E-1	.3075 E-1
5.00	0.4152	29.0000	5.0000	5.8000	.6172 E-1	.3062 E-1
5.01	0.4151	29.1168	5.0033	5.8195	.6124 E-1	.3050 E-1
5.02	0.4149	29.2338	5.0066	5.8390	.6077 E-1	.3038 E-1
5.03	0.4148	29.3510	5.0099	5.8586	.6030 E-1	.3027 E-1
5.04	0.4147	29.4685	5.0132	5.8782	.5984 E-1	.3015 E-1
5.05	0.4145	29.5862	5.0165	5.8978	.5938 E-1	.3003 E-1
5.06	0.4144	29.7042	5.0197	5.9175	.5893 E-1	.2991 E-1
5.07	0.4142	29.8224	5.0230	5.9372	.5848 E-1	.2980 E-1
5.08	0.4141	29.9408	5.0262	5.9570	.5803 E-1	.2968 E-1
5.09	0.4140	30.0594	5.0294	5.9768	.5759 E-1	.2957 E-1
5.10	0.4138	30.1783	5.0326	5.9966	.5715 E-1	.2945 E-1

(*Source:* Bush, M.B., Normal Shock Tables. Retrieved January 7, 2023, from https://web.iitd.ac.in/~pmvs/courses/mcl341/normal_shock.pdf)